SpringerBriefs in Molecular Science

SpringerBriefs in Molecular Science present concise summaries of cutting-edge research and practical applications across a wide spectrum of fields centered around chemistry. Featuring compact volumes of 50 to 125 pages, the series covers a range of content from professional to academic. Typical topics might include:

- A timely report of state-of-the-art analytical techniques
- A bridge between new research results, as published in journal articles, and a contextual literature review
- A snapshot of a hot or emerging topic
- An in-depth case study
- A presentation of core concepts that students must understand in order to make independent contributions

Briefs allow authors to present their ideas and readers to absorb them with minimal time investment. Briefs will be published as part of Springer's eBook collection, with millions of users worldwide. In addition, Briefs will be available for individual print and electronic purchase. Briefs are characterized by fast, global electronic dissemination, standard publishing contracts, easy-to-use manuscript preparation and formatting guidelines, and expedited production schedules. Both solicited and unsolicited manuscripts are considered for publication in this series.

Abir B. Majumder · Kalluri V. S. Ranganath

Understanding Kinetic Resolution by Hydrolases

Maximizing Enantioselectivity

 Springer

Abir B. Majumder
Department of Chemistry
KGT Mahavidyalaya
University of North Bengal
Siliguri, West Bengal, India

Kalluri V. S. Ranganath
Department of Chemistry
Institute of Science
Banaras Hindu University
Varanasi, Uttar Pradesh, India

ISSN 2191-5407   ISSN 2191-5415   (electronic)
SpringerBriefs in Molecular Science
ISBN 978-3-031-46352-5   ISBN 978-3-031-46353-2   (eBook)
https://doi.org/10.1007/978-3-031-46353-2

This Springer imprint is published by the registered company Springer Nature Switzerland AG
The registered company address is: Gewerbestrasse 11, 6330 Cham, Switzerland

Paper in this product is recyclable.

Dedicated to all budding scholars who want to taste the exquisiteness of kinetic resolution for targeted synthesis using hydrolases.

Abir B. Majumder
Kalluri V. S. Ranganath

Preface

Applied biocatalysis has gained enormous popularity in recent years for its "Green" nature in synthetic organic chemistry. Scientists have explored enzymes' high substrate specificity in gaining high output stereoselectivity. Among the different classes of enzymes hydrolases, especially serine hydrolases, have been randomly used for this purpose. The wide applicability of these hydrolases has made these as one of the most important class of enzymes that are frequently chosen in industries for chiral synthesis. One commonly used approach in industrial stereoselective syntheses is enzymatic kinetic resolution. The most important aspect of enzymatic kinetic resolution is to tune and optimize the process to get maximum possible enantiomeric excess in the product or in the unreacted starting material. While fast reactions are mostly demanded, a slower reaction may prove very encouraging in such resolution. Hence one must have a clear idea about the reaction, to know when he needs to slow down the process or compromise with the yield or conversion with enantiomeric excess and to quench it. An attempt has been made in this book to discuss briefly, not going much to the depth, to touch the various aspects of enzymatic kinetic resolution including the use of organometallic catalysts for dynamic kinetic resolution. A beginner, with a preliminary knowledge in enzymology, stereochemistry and organometallics, who wishes to take the challenge to resolve a racemic mixture using biocatalysis, on the bench, if finds this book handy, the purpose of this write up would be served.

Siliguri, India   Abir B. Majumder
Varanasi, India   Kalluri V. S. Ranganath

Acknowledgements

I wish to thank my parents, my wife Gopa and my son Abhilash for their continuous support to see this book in the present form. I am indebted to Prof. Munishwar N. Gupta who taught me how to use enzymes for organic synthesis and for his inspiring words that immensely helped me to fight against my adverse health and continue writing this. I would also like to thank Prof. Kalluri V. S. Ranganath who agreed to contribute as a co-author and has shown extreme patience in shaping it up.
Finally I would like to thank the following publishing houses (in alphabetical order) for the copyright clearance for reusing excerpts.
American Chemical Society
Elsevier Science and Technology Journals
Taylor & Francis Informa UK Ltd.—Journals

Abir B. Majumder
Department of Chemistry
KGT Mahavidyalaya
University of North Bengal
Siliguri, West Bengal, India

Contents

Abbreviations

BCL	*Burkholderia cepacia* lipase
c	Centigrade
C	Conversion
CALB	*Candida antarctica* lipase B
CLEA	Cross-linked enzyme aggregate
CLEC	Cross-linked enzyme crystals
E	Enantioselectivity
ee	Enantiomeric excess
ee_P	Enantiomeric excess of the product
ee_S	Enantiomeric excess of the substrate
enz	Enzyme
$G^{\ddagger}$	Gibb's free energy related to the transition state
G	Gibb's free energy
GC	Gas chromatography
H	Enthalpy
$H^{\ddagger}$	Enthalpy related to the transition state
HPLC	High performance liquid chromatography
k_{Cat}	Catalytic constant
k_M	Michaelis constant
K	Kelvin (unit of scale in absolute temperature)
k	Rate constant
NMR	Nuclear magnetic resonance
PPL	Porcine pancreatic lipase
RI	Refractive index
S	Entropy
$S^{\ddagger}$	Entropy related to the transition state
SEM	Scanning electron microscopy
T	Temperature (in K)
tert	Tertiary

v	Rate of a reaction
ε	Expression/equation
Journal abbreviations	As per International Standard Serial Number, www.issn.org.

Chapter 1
Understanding Enzymatic Kinetic Resolution

Abstract Kinetic resolution involves conversion of one enantiomer more readily to product than the other of the racemic mixture. Desymmetrization, on the other hand, indicates the destruction of the plane of symmetry of the starting reactant molecule by regioselective functionalization. Both the processes look very close by nature when the reaction goes but actually are widely different. Present chapter addresses the fundamentals of kinetic resolution aided by hydrolases, underlying mechanism and its difference with enzymatic desymmetrization.

Keywords Enzymatic kinetic resolution · Desymmetrization · Hydrolases · Transacetylation · Dynamic kinetic resolution

1.1 Enzymatic Kinetic Resolution (EKR): Importance

Enzymatic kinetic resolution (EKR) has gained enormous popularity in recent days for its high acceptability in sustainable chemistry [1]. Among the different classes of enzymes used frequently for the purpose, the use of hydrolases, especially that of lipases and proteases, has been widespread. Kinetic resolutions are more frequent because (out of combinatorial considerations) there are more racemic molecules in nature than *meso* and prochiral compounds. Consequently, racemates always have been (and will be) an indispensable starting point for the synthesis of chiral materials in non-racemic form for their ease of availability. In principle, kinetic resolution of racemates is based on the difference in reaction rates of enantiomers in the presence of a biocatalyst. In an EKR, the reactive "fitting" enantiomer is quickly converted, while the "wrong" enantiomer remains relatively untouched. Thus, in an ideal situation, the reaction comes to a standstill at 50% conversion, where both enantiomeric substrate and product can be separated by physical means.

1.2 Negativities of Enzymatic Kinetic Resolutions

Despite its widespread applications, kinetic resolution is associated with several inherent disadvantages, especially on an industrial scale. The most obvious drawbacks of kinetic resolution are:

- The theoretical yield of an enantiomer can never exceed a limit of 50%.
- Separation of the formed product from the remaining substrate may be laborious in particular for cases in which simple extraction or distillation fails and chromatographic methods are required.
- In most of the cases, only one stereoisomer (as biologically more active) is desired and the other may not be so useful. In some rare cases, the unwanted isomer may be used through a separate synthetic pathway which requires highly flexible enantioconvergent strategies [1, 2].

For kinetic reasons, the optical purity of the substrate or the product is depleted at the point where separation of product and substrate is most desirable from a preparative point of view, i.e., at 50% conversion [2]. However, the dynamic kinetic resolution, where the less reactive enantiomer is converted to the other enantiomer by chemical means during the kinetic resolution, if possible, provides a solution. We will come back to this in Chap. 4 later in this book. For this moment, without opening too much buckets, let us concentrate upon non-dynamic kinetic resolutions only.

1.3 Enzymatic Kinetic Resolution: Understanding the Process

The best way to understand the process of enzymatic kinetic resolution is to start with suitable illustrations (Schemes 1.1–1.3). A typical example of kinetic resolution of racemic 1-(4-methoxyphenyl)-ethanol via enantioselective transacetylation is given in Scheme 1.1. Lipase prefers (*R*)-alcohol, **1a**, and thus forms the acetate **2a** in preference to **2b** of the two possible enantiomeric products shown within the rectangular box in Scheme 1.1 [see Kazlauskas' empirical rule: Sect. 3.4]. At any instant of time, during the course of the reaction, concentration of **1b** would be higher than **1a** and that of **2a** would be higher than **2b**. Ideally, for a highly enantioselective synthesis, with the same model, reaction would cease at 50% conversion and a mixture of **2a** and **1b** (shown within the oval box in Scheme 1.1) would be obtained at the end.

A second type of kinetic resolution (not starting with racemic mixture) deals in the destruction of the plane of symmetry (the presence of a plane of symmetry makes the molecule optically inactive) of a single molecule by selective functionalization (Schemes 1.4 and 1.5). The process is known as desymmetrization. Thus, one of the acetyl groups of a symmetric prochiral (tetrahedral carbon bearing two identical functional groups and two different groups) compound 3 (Scheme 1.4), and *meso* diacetate 5 (Scheme 1.5 compound containing two identical chiral centers of

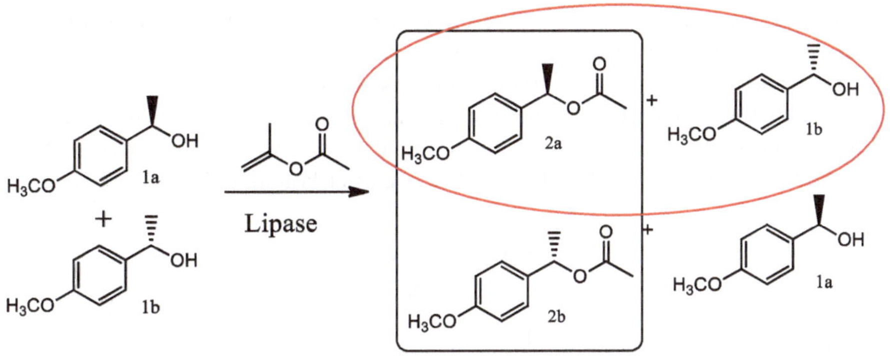

Scheme 1.1 Kinetic resolution of a racemic secondary alcohol: enantioselective transacetylation of 1-(4-methoxyphenyl)-ethanol [5, 6]

opposite configuration and bearing a vertical plane of symmetry or mirror in between them) is selectively hydrolyzed by lipases from *Candida rugosa* [3] and a lipase from *Rhizopus delemar* (Lipase D, Amano) [4], respectively.

In both the syntheses, the process is an enzymatic transacylation generically known as transesterification. The most common type of transesterification is also known as alcoholysis as aided by alcohols. This is not to be confused with interesterification where both the reactants are carboxylic acid esters and an interchange of acyl group takes place (Scheme 1.2). Out of these two reactions, transacylation by hydrolases has gained enormous popularity in stereoselective reactions [7].

In a transacylation process (Scheme 1.2), the acyl group RCO migrates from an ester, $RCOOR^1$ (in blue) to an alcohol (R^2OH in red). The process is also termed as alcoholysis. It would be good at this point if we try to understand the process

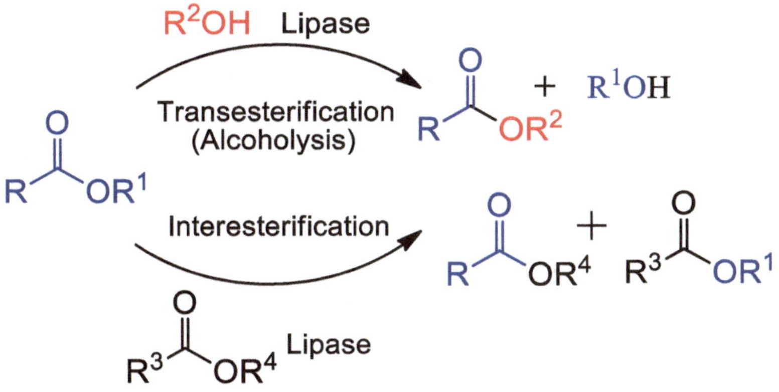

Scheme 1.2 Typical transacylation reaction catalyzed by lipases

mechanistically as well. Scheme 1.3 describes a typical transacylation (or alcoholysis of an ester) reaction catalyzed by serine hydrolases [8].

The serine residue, of the catalytic triad, acts as a nucleophile and binds with the ester in step **A**, forming a tetrahedral intermediate (II, Scheme 1.3) which is stabilized by hydrogen bonding interactions. In the next step **B**, acylation of serine OH is completed and the alcohol, R_1OH, which was earlier bound to the ester liberates.

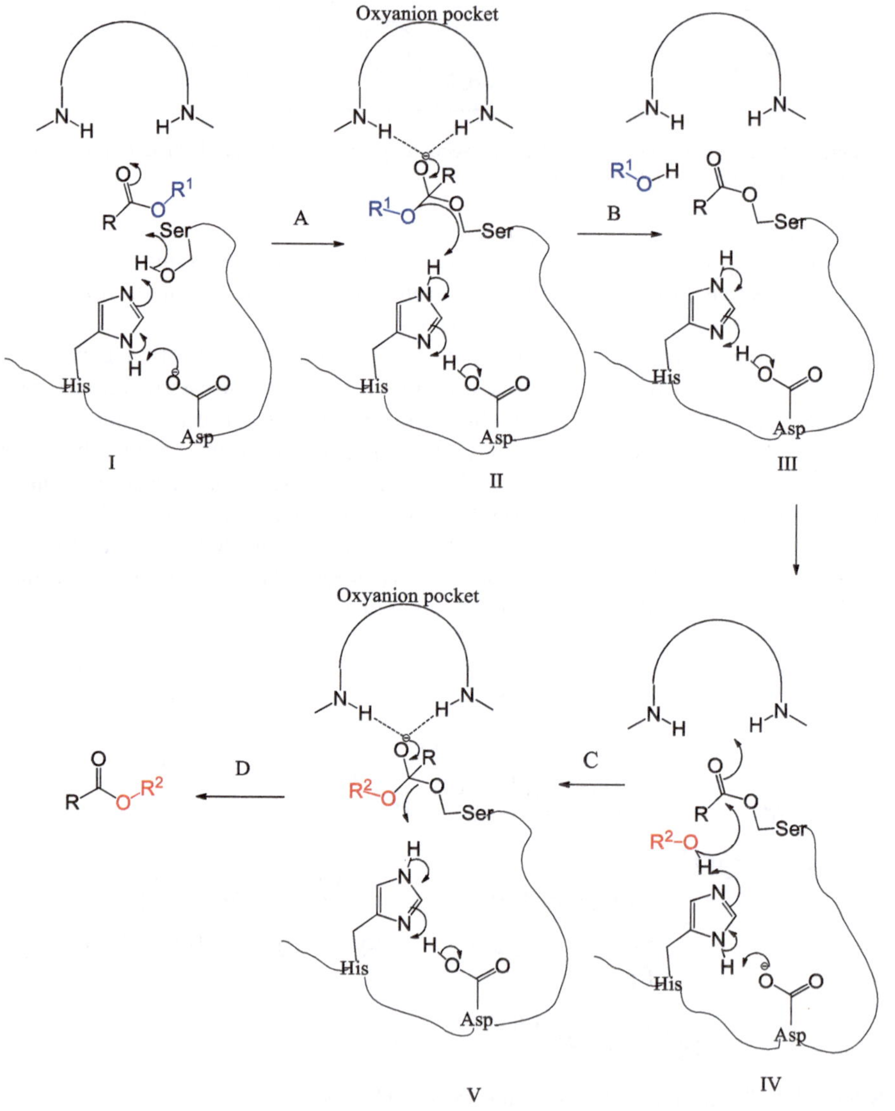

Scheme 1.3 Mechanistic details of enzymatic transacylation reactions catalyzed by serine hydrolases

This step (B, Scheme 1.3) is very significant. The liberation of alcohol from an ester in a transacylation process is often taken as due to hydrolysis! In the presence of water, viz. in biphasic systems containing aqueous-organic media or reactions in organic solvents saturated with water, there are parallel processes which may cause a fraction of alcohol R^1OH liberated due to hydrolysis as well [9], but in low-water media where the water content is just enough to hydrate the enzyme (for its optimum flexibility), the presence of R^1OH in the reaction medium cannot be taken as due to hydrolysis. However, in a recent study it has been stated that some hydrolases may act as acyltransferases even in the presence of water, and the rate of transacylation may be higher than that of hydrolysis. The reason behind this is not clear [10].

A more appropriate term to describe this detachment of acyl group from R^1OH in non-aqueous media thus seems to be "deacylation". This has been justified from the fact, that in nearly anhydrous media, while performing transacylation with vinyl acetate, acetaldehyde could be detected in high concentration and was found to be responsible for many unusual reactions [10, 11]. The rate of the formation of acetaldehyde was too high to justify its availability because of hydrolysis, for the concentration of water in the media was "close to zero".

Finally, the second alcohol molecule (R^2OH) attacks as a nucleophile and a second tetrahedral intermediate is formed through step C [9B] which is more significant for the enantioselectivity (E, described in Sect. 2.1.3, Chap. 2) of the enzymatic reaction as it contains the enantiomeric alcohol (R^2OH, in red, Scheme 1.3). This is the step where the enzyme picks up an enantiomer of the racemic alcohol preferentially over the other. This step seems to be the slowest step during the entire process and hence rate determining, for the enantiomeric substrate (to be preferentially accepted) requires a number of favorable interactions to be developed through the formation of a transition state and then the intermediate through a specific suitable orientation of the acylated enzyme–enantiomeric substrate alcohol pair. The rate constant of this step (C, Scheme 1.3), k_{cat}, thus gets significance in calculation of enantioselectivity described in section in Chap. 2. An extremely fast deacylation (steps A, B, Scheme 1.3) through the formation of the first tetrahedral intermediate is observed with vinyl esters (as acyl donors, Sect. 3.1.5.1, Fig. 3.1). Interestingly, when a racemic ester is the substrate, and we are looking for an enantiopure unreacted ester or product ester (or product acid) and R^2OH is achiral, both the steps A and C where the tetrahedral intermediates are formed involving the chiral substrate contribute to overall k_{cat} and hence find importance in shaping enantioselectivity. Finally, step C is followed by a fast release of the product molecule $RCOOR^2$. For these two steps, steps C and D, the process is also termed as "alcoholysis".

The readers must note that when we mention hydrolysis of an ester by a serine hydrolase, these last two steps involve the attack of a water molecule instead of an alcohol; i.e., R^2OH must be HOH. Coming back to desymmetrization, the enzymatic transacylation destroyed the vertical plane of symmetry (in Schemes 1.4 and 1.5, shown by the broken line) in the molecules (3, Scheme 1.4; 5, Scheme 1.5) resulting in products which are hence optically active.

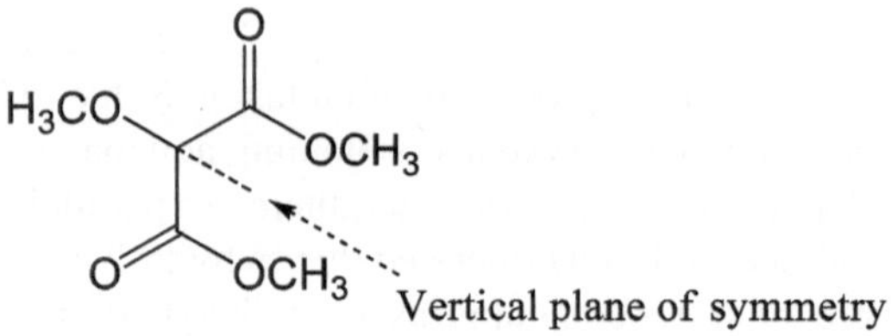

3

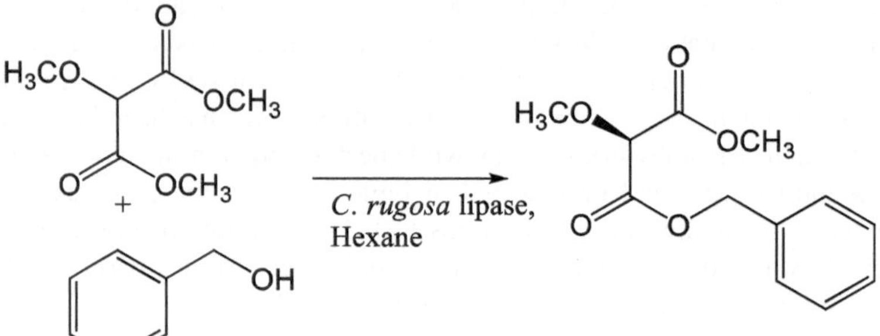

4

98 % ee, 53 % conversion

Scheme 1.4 Desymmetrization of a monosubstituted prochiral malonate [3]

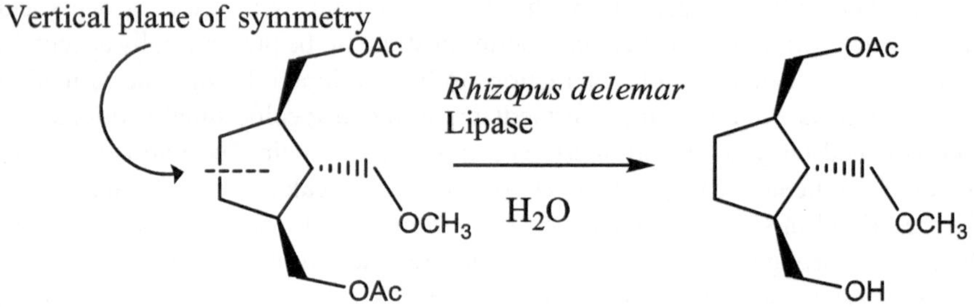

Yield 43 %, 99 % ee

5

6

Scheme 1.5 Desymmetrization of a *meso* diacetate by stereoselective hydrolysis catalyzed by Lipase® D, Amano [4]

Kinetic resolution and desymmetrization apparently may look widely different, but these two processes do follow a similar course of the reaction. The following section addresses this topic.

1.4 Kinetic Resolution and Desymmetrization

Kinetic resolution always starts with a pair of enantiomers: molecules which are non-superimposable mirror images of each other. An enzyme picks up one, as it fits in its active site better and catalyzes the reaction leaving an "enantioenriched (in the misfit)" unreacted substrate concentrate.

Desymmetrization, on the other hand, always starts with a single molecule. Interestingly, both the processes in several reports gave the best results, i.e., the highest possible enantiomeric excess, near about 50% conversion. An obvious question comes in mind, why? To understand it, a better term "functionalization" instead of "conversion" should be used for desymmetrization processes. It should be borne in mind that here one symmetric molecule provides two functional groups (on opposite side of the mirror plane) and hence giving the effect of two substrate molecules and acting as a conjugated pair of enantiomers. Thus at 50% conversion for a highly enantioselective biocatalysis, one may expect half of the substrate molecules (for a racemic mixture) or half of the functional groups (for desymmetrization) are converted.

The fact that why does a stereoselective biocatalysis dealing with an enzymatic kinetic resolution give its best output near 50% conversion will be discussed in the upcoming chapter. For the time moment, the need is to look for the justification that helps us to understand that both the processes are logically close. Well, to satisfy ourselves to an extent, for desymmetrization one can visualize the symmetric molecule as both the enantiomers joined together via a plane of symmetry. Hence, a suitable biocatalyst finds a symmetric molecule more or less like a pair of enantiomers joined through a non-reactive part (which has nothing to do with enzyme active site) and treating it as a racemic mixture fits it in such a way that one side of the mirror reacts and the other half remains as a misfit and hence unreacted. In the next chapter, the various components of an enzymatic kinetic resolutions are discussed.

References

1. (A) Sheldon RA, Woodley JM (2018) Role of biocatalysis in sustainable chemistry. Chem Rev 118(2):801–838. (B) Wu S, Snajdrova R, Moore JC, Baldenius K, Bornscheuer UT (2021) Biocatalysis: enzymatic synthesis for industrial applications. Angew Chem Int 60:88–119. (C) Patel RN (2000) Stereoselective biocatalysis. Mercel–Dekker, New York. (D) Schulze B, Wubbolts MG (1999) Biocatalysis for industrial production of fine chemicals. Curr Opin Biotechnol 10:609–615
2. (A) Cotterill IC, Sutherland AG, Roberts SM, Grobbauer R, Spreitz J, Faber K (1991) Enzymatic resolution of sterically demanding bicyclo[3.2.0]heptanes: evidence for a novel hydrolase in crude porcine pancreatic lipase and the advantages of using organic media for some of the biotransformations. J ChemSoc Perkin Trans 1:1365–1368. (B) Gupta PK, Kumar N, Majumder AB, Pandey M, Goverdhan RPV, Ranganath KVS (2023) Chiral modification of Ferrite Nanoparticles for oxidative kinetic resolution of Benzoins. Asian J Org Chem 12(8): e202300325 (1–6)
3. Chen C-S, Fujimoto Y, Girdalukas G, Sih CJ (1982) Quantitative analyses of biochemical kinetic resolutions of enantiomers. J Am Chem Soc 104:7294–7299

4. Gutman AL, Shapira M, Boltanski A (1992) Enzyme-catalyzed formation of chiral mono-substituted mixed diesters and half esters of malonic acid in organic solvents. JOrgChem 57:1063–1065
5. Ghanem A, Schurig V (2003) Entrapment of *Pseudomonas cepacia* lipase with peracetylated β-cyclodextrin in sol–gel: application to the kinetic resolution of secondary alcohols. Tetrahedron Asymm 14:2547–2555
6. Ghanem A, Aboul-Enein H (2005) Application of lipases in kinetic resolution of racemates. Chirality 17:1–15
7. Sakai K, Tanaka M, Yoshioka M (1993) Highly asymmetric enzymatic hydrolysis and transesterification of *meso*-biscacetoxymethyl)- and bis(hydroxymethyl)cyclopentane derivatives: an insight into the active site model of *Rhizopus delemar* lipase. Tetrahedron Asymm 4:981–986
8. (A) Luić M, Tomić S, Ljubović E, Šepac D, Šunjić V, Vitale L, Saenger W, Kojić-Prodić B (2001) Complex of *Burkholderia cepacia* lipase with transition state analogue of 1-phenoxy-2-acetoxybutane: biocatalytic, structural and modelling study. Eur J Biochem 268(14):3964–3973. (B) Jiang Y, Morley KL, Schrag JD, Kazlauskas RJ (2011) Different active-site loop orientation in serine hydrolases versus acyltransferases. ChemBioChem 12:768–776
9. (A) Weber HK, Faber K (1997) Stabilization of lipases against deactivation by acetaldehyde formed in acyl transfer reactions. In: Rubin B, Dennis EA (eds) Methods in enzymology. Academic Press Inc., Elsevier, 286, pp 509–518. (B) Weber HK, Weber H, Kazlauskas RJ (1999) Watching lipase-catalyzed acylations using 1H NMR: competing hydrolysis of vinyl acetate in dry organic solvents. Tetrahedron Asymm 10:2635–2638
10. Müller H, Becker A-K, Palm GJ, Berndt L, Badenhorst CPS, Godehard SP, Reisky L, Lammers M, Bornscheuer UT (2020) Sequence-based prediction of promiscuous acyltransferase activity in hydrolases. Angew Chem Int Ed Engl 59(28):11607–11612
11. Majumder AB, Ramesh NG, Gupta MN (2009) A lipase catalyzed condensation reaction with a tricyclic diketone: yet another example of biocatalytic promiscuity. Tetrahedron Lett 50(37):5190–5193

Chapter 2
Essential Parameters: Determination and Significance

Abstract The most interesting and important aspect of stereoselective biocatalysis is to find out how selective the process is. The assessment of stereoselectivity exhibited by an enzyme in kinetic resolution is determined after finding out the values of enantiomeric excess of the products or that of the unreacted remaining substrate at a certain conversion value, and then enantioselectivity is calculated. This may also be calculated from the initial rates of the same reaction starting from two enantiomeric pure substrates (of the same racemic mixture) with the same enzyme formulation under the same reaction condition. This chapter describes the steps, in detail, to ascertain the value of enantioselectivity of an enzymatic process and its significance in industrial applications.

Keywords Enantiomeric excess · Optical rotation · Specific rotation · Chromatography · Enantioselectivity

2.1 Parameters of EKR

2.1.1 Enantiomeric Excess (ee)

Optical purities are usually expressed in terms of enantiomeric excess. During a kinetic resolution, for example in Scheme 1.1, Chap. 1, at any time instance, the % excess of 1b over 1a is termed as substrate enantiomeric excess (ee_S) and the % excess of 2a over 2b as the enantiomeric excess of the product (ee_P).

Enantiomeric excess is defined by the following expression [1].

$ee_S = \frac{[1b]-[1a]}{[1b]+[1a]}$, where [1a] and [1b] are the concentrations of enantiomers 1a and 1b. Similarly, the product enantiomeric excess, ee_P, would be $= \frac{[2a]-[2b]}{[2a]+[2b]}$.

The *ee* values are always expressed in %; so the *ee* values in fractions obtained by the above expressions are multiplied by 100.

© The Author(s), under exclusive license to Springer Nature Switzerland AG 2023
A. B. Majumder and K. V. S. Ranganath, *Understanding Kinetic Resolution by Hydrolases*, SpringerBriefs in Molecular Science, https://doi.org/10.1007/978-3-031-46353-2_2

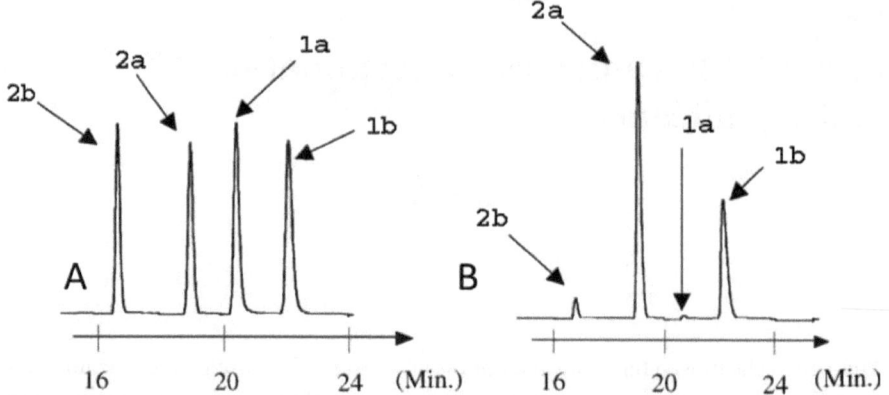

Fig. 2.1 Chiral GC analysis of enantioselective transacetylation of racemic 1-(4-mthoxy-phenyl) ethanol (Scheme 1.1) **A** chemical acetylation after 50% conversion **B** enzymatic transacetylation after 53% conversion [2]. Reused with permission from [3]. Copyright 2004, Elsevier Science & Technology Journals

2.1.2 Determination of ee

Enantiomeric excess values can be determined in the following ways.

2.1.2.1 By Chiral GC or HPLC Analysis

GC or HPLC fitted with chiral columns (also known as chiral GC or HPLC) can resolve the enantiomers. Thus, 1a, 1b, 2a, 2b (Scheme 1.1, Chap. 1) would give four different close peaks under optimized programming conditions in a GC fitted with "C-11 ChirasilDex" chiral column (Fig. 2.1). In such cases, *ee* values are directly obtained from the peak areas of the enantiomers using the expressions given above [2]. Here the peak areas can be used directly for the concentrations. It is to be mentioned that chiral columns are very specific and one column which can resolve a racemic mixture may not be suitable at all for another one, and for this latter pair, it would give a single peak.

2.1.2.2 Using Polarimetry

In cases where the chiral GC or HPLC fails to separate the enantiomeric peaks, *ee* can be determined conveniently by polarimetric measurements or by ^{1}H-NMR spectra using chiral shift reagents. However, the most sensitive and convenient methods used are by chiral GC or chiral HPLC [3]. To measure *ee* by polarimetry, for example in the above reaction, a workup is required to separate and purify the product mixture of 2a and 2b from the unreacted substrate mixture containing 1a and 1b. This purification step is very important to make sure that there is no contribution of optically active (or inactive) impurities on the optical rotation; even if the impurity is optically inactive, it would give rise to an increased concentration value and accordingly a decreased specific rotation.

Hence, before doing polarimetry it is always better to check the purity of the compound mixture by ^{1}H-NMR (or by HPLC using RI detector). Thus, in this process, *ee* values of the unreacted substrate mixture and of the product mixture are determined in separate experiments. The optical rotation or the specific rotation obtained in these cases is actually the subtracted value of the rotations exhibited by the two enantiomers (present in different amounts), and if this value for example is $[\alpha]_{mix}$, the *ee* would be $= \frac{[\alpha]_{mix}}{[\alpha]_{pure}} \times 100$, where $[\alpha]_{pure}$ is the specific rotation of the optically pure enantiomer under the same experimental condition. Thus, the value of $[\alpha]_{pure}$ must be known.

In this context, the most relevant but fundamental question may pop: What is the relationship of optical rotation with the plane-polarized light and with the structure of a molecule? **How does a chiral molecule show rotation in a polarimeter**?

The classical electromagnetic theory assumes a molecule to be mere a combination of electrons and nuclei. When a molecule possessing chirality interacts with the electromagnetic radiation, it accelerates in a partially circular path and generates a magnetic force. When this force interacts with the applied electromagnetic field, it shifts its electric vector away from the polarization plane. Fresnel discovered the sinusoidal nature of the electric component of linear/plane-polarized light. When two circularly polarized beams travel through the same media, it meets a certain non-uniform distribution of charged particles and its interaction affects molecule velocity through the medium. For clockwise polarization, the interaction and velocity both will be different from anticlockwise polarization. Therefore, the speed of one of the circularly polarized components increases or decreases relative to the speed of the other. This is easier to visualize if one thinks about a circularly polarized beam interacting with a right-handed or left-handed helix very much like a right-handed screw that moves easily through a thread made for it assisting its movement but unable to move forward through a left-handed thread. When both clockwise and anticlockwise components are present in the same sample, the plane of linear polarization will be rotated either toward left or right direction. A cumulated contribution results in the rotation in either direction. However in case of achiral molecules, the plane of symmetry, and hence no change of velocity of one circularly polarized beam relative to another circularly polarized beam, thereby causing no rotation of the exiting recombined linearly polarized light. Biot's convention of assigning (+) and (−) prefixes for clockwise (dextrorotatory) and counterclockwise (levorotatory) rotation of the plane of polarization with respect to the observer has made it easy to correlate molecule's chirality with light. For a routine analysis, a standard light source of wavelength ($\lambda =$ 589.3 nm; sodium lamp) is used for analyzing the optical purity of the sample (only when it has zero not absorbance this region). The specific rotation is determined using standard cell length and concentration through the following equation:

$$[\alpha]_{589.3}^{T} = \frac{\alpha}{l.c},$$

where α, T, l and c refer to rotation (degrees), temperature (K), path length (dm) and concentration (g/mL) or density (for pure sample).

For example, specific rotation of a standard sample of (S)-1-phenylethanol has been reported as $-$ 31.2 (in methanol). It should be noted that the sample and/or solvent must be sufficiently pure. Addition of small impurities can change its value significantly and may give weird result. For example, while dealing with oxidative kinetic resolution of racemic 1-phenylethanol, during polarimetric studies, as a control experiment, addition of a drop of acetophenone (in methanol) of same concentration to the cell has been found to change specific rotation of pure (S)-1-phenylethanol to $-$ 70.3 (in methanol). This shows that the presence of even a trace amount of an optically inactive but structurally close molecule (or impurity) may change the specific rotation drastically.

2.1.3   Enantioselectivity, E: Concepts in Biocatalysis

Enantioselectivity, the term which was first introduced by Chen et al. [1], has become a very important parameter in enzymatic kinetic resolution. It virtually reflects the efficiency of an enzyme formulation to resolve an enantiomeric pair for a particular reaction.

Let, "a" and "b" be the fast and the slow reacting substrate enantiomers (as in Scheme 1.1, Fig. 2.1), respectively, which are competing for the same active site of the enzyme.

For this, a simple three-step kinetic mechanism would be like Scheme 2.1. Thus, when in a kinetic resolution the unreacted substrate is of greater importance, the target should be to not only get an $ee > 90\%$ and but with a minimum conversion to the product so that the yield of the enantiopure substrate is high.

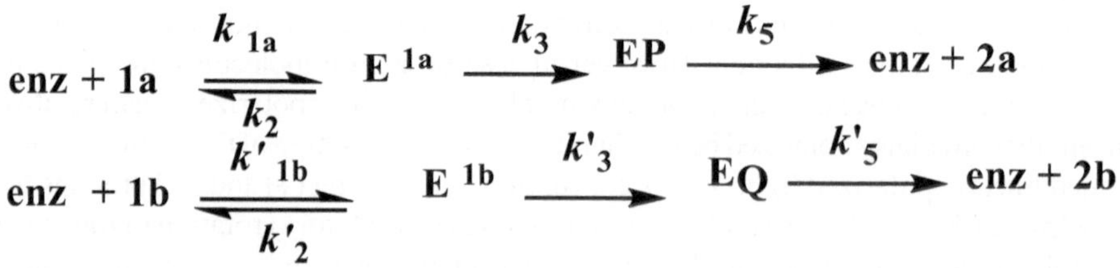

$$\text{enz} + \text{1a} \; \underset{k_2}{\overset{k_{1a}}{\rightleftharpoons}} \; \text{E}^{\,1a} \; \xrightarrow{k_3} \; \text{EP} \; \xrightarrow{k_5} \; \text{enz} + \text{2a}$$

$$\text{enz} + \text{1b} \; \underset{k'_2}{\overset{k'_{1b}}{\rightleftharpoons}} \; \text{E}^{\,1b} \; \xrightarrow{k'_3} \; \text{EQ} \; \xrightarrow{k'_5} \; \text{enz} + \text{2b}$$

Scheme 2.1 EKR: The model 1a and 1b are the two substrate enantiomers as described in Scheme 1.1, Chap. 1; enz stands for enzyme. Adapted with permission from [1]. Copyright 1982, American Chemical Society

"E^{1a} and E^{1b}" are the enzyme bound 1a and 1b complexes, respectively. Assuming the reaction is irreversible and there is no inhibition as in Chen's model, the rate of formation of 2a from enantiomer 1a (following Michaelis–Menten kinetics) would be $= \frac{V_{max}[1a]}{k_{Ma}}$, (where and that for 2b from 1b $= \frac{V_{max}[1b]}{k_{Mb}}$ and the ratio of the two rates:

$$\frac{v_{1a}}{v_{1b}} = E = \frac{(V_{max}/k_M)a}{(V_{max}/k_M)b}$$
$$= \frac{(k_{cat} \times [E]_o/k_M)a}{(k_{cat} \times [E]_o/k_M)b}$$
$$= \frac{(k_{cat}/k_M)a}{(k_{cat}/k_M)b}. \tag{2.1}$$

It is to be noted that at the start of the reaction, for a perfect racemic mixture, [1a] = [1b]. This ratio of the two rates is termed as the **enantiomeric ratio** or **enantioselectivity, E**. Obviously, this expression can be used if and only if the reaction is started separately with enantiomer "a" and enantiomer "b".

Unfortunately, kinetic resolution starts with a racemic mixture, and the mathematically deduced alternative equations are hence more popular which involve *ee* and conversion values.

$$E = \frac{Ln[1 - c(1 + ee_P)]}{Ln[1 - c(1 - ee_P)]} \tag{2.2}$$

$$= \frac{Ln[(1 - c)(1 - ee_S)]}{Ln[(1 - c)(1 + ee_S)]} \tag{2.3}$$

Here, "C" is the degree of conversion which is calculated using the following expression:

$$C = \frac{ee_S}{(ee_S + ee_P)} \tag{2.4}$$

This value, practically on the bench, only fractionally differs from the value obtained in a usual way, i.e., using the following expression:

$$\text{Conversion} = \frac{\text{Product concentration}}{\text{(Unreacted substrate} + \text{Product) concentration}}$$
$$= \frac{[2a] + [2b]}{[1a] + [1b] + [2a] + [2b]}.$$

In general, for kinetic resolution, conversion would mean the "degree of conversion" being calculated by using expression (2.4). Practically both conversion and the degree of conversion may differ negligibly in values but would have a significant impact on enantioselectivity.

2.2   Enantioselectivity, Enantiomeric Excess and Conversion During the Course of an EKR

2.2.1   *Dependence of the Substrate ee on Conversion*

Figure 2.2 shows the variation of enantiomeric excess of the unreacted substrate, ee_S during the course of the five different reactions **A1-5** (or five different conditions of the same reaction as each of these would correspond to a different E). All of these can give rise to an $ee_S > 95\%$ but only at different conversions.

Taking this into consideration, the most acceptable condition would be **A5** which gives ee_S of 1b > 90% at around 50% conversion (maximum yield = 50%). For **A4** to reach the same level of enantiopurity, we need to go around 55% conversion (maximum yield of the unreacted substrate = 45%), and this value goes on increasing from **A3 to A1**. Thus in the case of **A1**, to get an ee even 80% one need to reach 80% conversion which would give a maximum of 20% yield of the unreacted substrate.

2.2.2   *"E is More Significant than ee"*

Thus, it is now clear that ee is not a sufficient data until the corresponding conversion value is known. If we observe the enantioselectivity values of the reactions (or reaction conditions), the best value is obtained in the case of **A5**. **Thus, enantioselectivity is a more useful parameter (than *ee*) which incorporates both the**

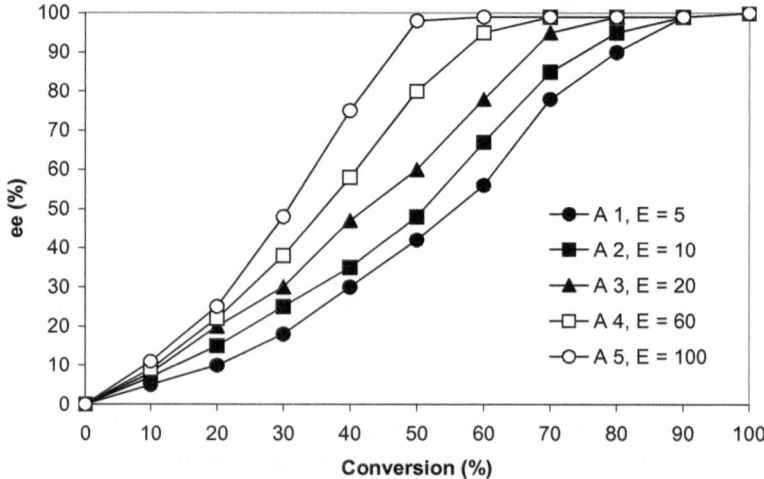

Fig. 2.2 Change in substrate *ee* with conversion. Adapted with permission from [1]. Copyright 1982, American Chemical Society

conversion and *ee*. As mentioned in Eq. (2.1), it signifies a ratio of the initial rates of the fast reacting enantiomer to the slow reacting enantiomer, and hence, the *E* **value** (italicized *E* stands for enantioselectivity), **for a single biotransformation, remains constant**. For an enantioselective biotransformation, the higher the *E* value, the more efficient is the kinetic resolution and so also is the industrial acceptability.

2.2.3 Dependence of the Product ee on Conversion

As in the case of the substrate, *ee*$_P$ value varies with conversion, though the pattern is not same. In this case, if the interest is in the product, the target would be to achieve a probable maximum *ee*$_P$ with a better conversion value, i.e., a higher product concentration or yield.

Figure 2.3 shows the variation of the *ee*$_P$ with conversion. In this case also, the best condition is **B5** which results in > 95% *ee*$_P$ at 55% conversion. It should be noted that *ee*$_P$ starts with a maximum value which then gradually decreases, and after 50% conversion, this decrease is sharp, more in the cases with higher *E*. The reason behind this is that, after 50% conversion in a highly enantioselective biotransformation, the concentration of the preferred enantiomer decreases more rapidly as compared to the other enantiomer and enzyme starts to pick this one (if it has to) having no other option.

Thus if one is interested into the product, the reaction should be stopped around 50% conversion which would give a more enantiopure product than at higher conversion values.

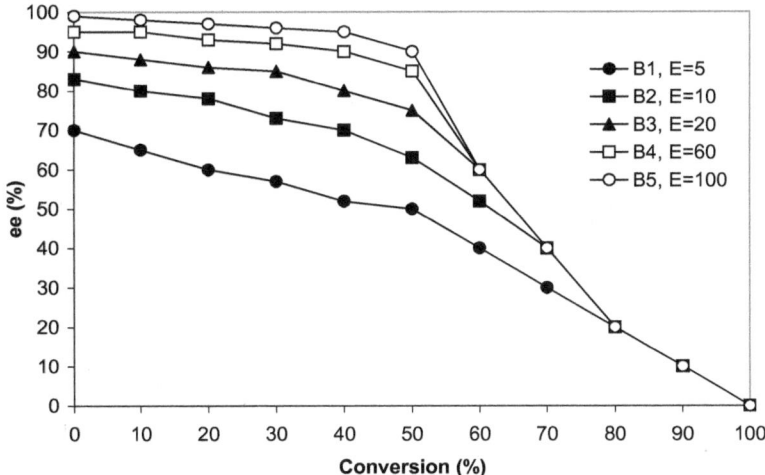

Fig. 2.3 Change in product *ee* with conversion. Adapted with permission from [1]. Copyright 1982, American Chemical Society

2.3   Condition for a High Enantioselectivity

In an ideal situation, when the E value is exceedingly high, the reaction would stop at or would be very slow after 50% conversion [1, 4]. Going back to Figs. 2.2 and 2.3, i.e., viewing **A5** and **B5** together ($E = 100$), the picture becomes more clear. From the figures, it is evident that with the increase in conversion (and time) ee_S increases sharply and reaches 90–95% at around 50% conversion and ee_P decreases only very slightly till 50% conversion where it also nears 90–95% value.

Thus around 50% conversion, ee_S and ee_P become numerically close which can also be understood from the expression (2.4) which gives a value of 0.5 when $ee_S = ee_P$. It has already been mentioned that a single reaction can give different E values under different conditions. In the next chapter, the experimental factors on which E depends are discussed.

References

1. Chen C-S, Fujimoto Y, Girdalukas G, Sih CJ (1982) Quantitative analyses of biochemical kinetic resolutions of enantiomers. J Am Chem Soc 104:7294–7299
2. Ghanem A, Schurig V (2003) Entrapment of Pseudomonas cepacia lipase with peracetylated β-cyclodextrin in sol–gel: application to the kinetic resolution of secondary alcohols. Tetrahedron Asymm 14:2547–2555
3. Ghanem A, Aboul-Enein H (2004) Lipase mediated chiral resolution of racemates in organic solvents. Tetrahedron Asymm 15:3331–3351
4. Bornscheuer UT (2000) Strategies for improving lipase-catalyzed preparation of Chiral compounds. In: Gupta MN (ed) Methods in non-aqueous enzymology. Birkhauser Verlag, Basel, Switzerland, pp 90–109

Chapter 3
Enantioselectivity: The Decisive Factors

Abstract The enantioselectivity of an enzymatic process has been found to depend upon many experimental factors. Effectiveness of a stereoselective biocatalysis depends upon how efficiently these conditions are tuned to achieve a targeted result. This chapter talks about these important factors with thermodynamic interpretation wherever needed and strategies to obtain a high E value.

Keywords Enantioselectivity · Water activity · Temperature · Acyl donor · Cross-linked enzyme aggregates

3.1 Factors Affecting Enantioselectivity

It has been pointed out earlier that "enantioselectivity" E itself is a property (for a particular reaction), exhibited by an enzyme, that depends upon the enantiomeric excess, *ee*, values and the extent of the conversion to the product, C. In the following sections, the factors that affect either *ee* or the C are discussed. Any change in either of these two or in both of these parameters would bring about a change in the E of the process.

3.1.1 Effect of Temperature

Temperature of a reaction is a very important parameter which controls the enantioselectivity of an enzyme. In an enzymatic reaction, the enantioselectivity is temperature dependent and obeys the following thermodynamic equation [1]:

$$\text{Ln } E = \Delta G^{\ddagger} = \frac{\Delta \Delta S^{\ddagger}}{R} - \frac{\Delta \Delta H^{\ddagger}}{RT} \tag{3.1}$$

Thus according to Eq. (3.1), enantioselectivity, E, is not only the result of enthalpic energy differences ($\Delta \Delta H^{\ddagger}$) between enantiomers in their respective transition state

A. B. Majumder and K. V. S. Ranganath, *Understanding Kinetic Resolution by Hydrolases*, SpringerBriefs in Molecular Science, https://doi.org/10.1007/978-3-031-46353-2_3

‡ but also contains an entropic component ($\Delta\Delta S^{\ddagger}$). At a certain temperature, the enthalpic and entropic effects will cancel each other making Ln E zero and the E will be 1. This indicates that a racemic product will be likely to be formed at this temperature. This temperature is called the racemic temperature. Depending on if the reaction is carried out above or below the racemic temperature, a decrease in temperature will cause either a decrease or increase in enantioselectivity [2].

This group of researchers has shown that, in an esterification of 2-arylpropanoic acid with 1-heptanol, lipases from *Rhizomucor miehei*, *Rhizopus oryzae* and *Candida rugosa* behaved differently from the lipase B of *Candida antarctica*. With the first three enzymes, E values were found to increase with a decrease in temperature, while with CALB a decrease was observed with a decrease in temperature. In this regard, it was pointed out that whether an increase of E with an increase in temperature would occur depends upon a favorable combination of transition state parameters which is not very uncommon. All that is required for this situation to appear is that $\Delta\Delta S^{\ddagger}$ and $\Delta\Delta H^{\ddagger}$ should have the same sign and that the absolute value of $\Delta\Delta H^{\ddagger}$ is smaller than that of $T\Delta S^{\ddagger}$. This has been found previously to be the case in some reactions catalyzed by alcohol dehydrogenases [2].

There are numerous cases observed earlier which showed that better enantioselective biotransformations were achieved at lower temperatures [3–5]. This is usually expected on the basis that at higher temperature, the enzyme becomes more flexible and can also accept the other enantiomer with a greater ease.

3.1.2   Effect of Water Content

As mentioned earlier, the presence of minimum amount of water is essential under anhydrous conditions for enzyme activity and it is even more critical for enantioselective synthesis. Addition of water decreases the intramolecular electrostatic interactions and thus increases the conformational flexibility of the enzyme which, beyond a certain limit, results in a more flexible enzyme conformation around the active site. Thus, the effect, to some extent, is similar as in the case of increase in temperature. Enantioselective biotransformations are preferred with lower water content with an optimally flexible enzyme formulation [6]. More flexibility often gives rise to a better initial rate (as both the enantiomers can be increasingly picked up) but a poor enantioselectivity as the difference in the initial rates corresponding to the fast reacting enantiomer and the slow reacting enantiomer decreases. A highly enantioselective transformation ($E > 100$) aided by the different enzymes with a same t system, viz. in enantioselective esterification of 2-octanol with decanoic acid was found to be only slightly affected by water activity (the fraction of water available to the enzyme surface), but it played a more important role in more hydrophobic solvents where the changes were more drastic [7].

Overall, when the reaction is highly enantioselective, the enzyme is compelled (by structural features) to accept only one enantiomer and the parameters which increase the initial rates also would likely to increase enantioselectivity.

However, the relationship between enantioselectivity and water activity (or water content) practically appears more complex. Table 3.1 reveals the unpredictability [8–15] of the change of E with a decreased water activity. Lowering of water content has been found to decrease E in cases [8, 13], while in majority of reports it has been found to increase [9, 12, 14, 15] or even without showing a marked difference [11].

3.1.3 Effect of the Presence of Water Miscible Organic Co-solvents

The presence of water miscible organic co-solvents like acetonitrile, dimethyl formamide and dimethyl sulfoxide has been found to be beneficial in some cases. However, no straightforward prediction can be done about the effect of the solvent on enantioselectivity, and in some cases, a complete reversal of E value was observed [16]. It was stated that co-solvents stabilize the diastereomeric transition states of the different enantiomers to a different extent resulting in a change in the k_{cat}/k_M values, and hence, a change in the enantiomeric ratio is observed. Thus, it is expected that the enzymes with a relatively "closed" active site, which cannot be accessed and hence stabilized or destabilized by the solvent molecules, would be indifferent to such change [17]. Thus, the lipase from porcine pancreas and the protease from α-chymotrypsin were found to be indifferent toward the effect of co-solvent [18].

In a reaction where water is generated as a side product (viz. esterification of alcohols), the presence of water miscible organic solvents removes the excess water molecules [19] which prevents the backward reaction and a consequent lowering of ee_P was observed.

For biocatalysis in the presence of a solvent without any co-solvent, more hydrophobicity was found to decrease enantioselectivity [20, 21].

3.1.4 Effect of Immobilization

Immobilization makes enzyme less flexible and hence results in a biocatalyst with better enantioselectivity. However, the improvement depends upon the way it has been immobilized [22]. In some cases, cross-linked lipases (both CLEC and CLEAs) have been recognized as excellent biocatalysts for enantioselective biotransformations [23]. Also non-covalent enzyme immobilization can often be found to improve enantioselectivity [24]. Recently, CLEAs in many cases have been reported to exhibit an increased enantioselectivity [24, 25]. The increase in E value was attributed due to partial purification and/or due to cross-linking which maintained the conformational stability in non-aqueous media.

Table 3.1 Change in *E* values at various water activities in lipase catalyzed kinetic resolutions

Entry	Presence of chiral center	*E* at lowered a w (or water content)	Hydrolase	Solvent	References
1		Lower	*C. cylindracea* lipase	Cyclohexane/ water	[8]
2	R*OH	Higher	*C. rugosa* lipase	Isooctane	[9]
3		Unchanged	*C. rugosa* lipase, *B. cepacia* lipase, *C. antarctica* lipase B	Hexane	[10]
4		Unchanged	*B. cepacia* lipase	Chloroform	[11]
5		Higher	*C. antarctica* lipase B	n-Hexane	[12]
6		Unchanged	*C. antarctica* lipase B	Dichloromethane	[12]
7		Lower	*C. rugosa* lipase	Cyclohexane	[13]
8		Higher	*B. cepacia* lipase	Dodecane	[14]
9		Higher	*Rhizomucor miehei lipase*	Solvent free (excess vinyl acetate)	[15]

3.1.5 Choice of the Acyl Donor: EKR via Transacylation

Chen's equation is based upon an irreversible enzymatic reaction model without any inhibition of the product. Thus, the parameters which make the reaction faster or irreversible would cause a better enantioselectivity. In this connection, it is to be noted that an increase in enzyme flexibility may help in achieving a better **overall** reaction rate, as it may accept both the enantiomers with a considerable decrease in E value. Furthermore, it is to be added that a minimum flexibility to make the enzyme functional is needed. But the flexibility we are talking about here is something more than what is needed, i.e., beyond the layer of hydration. Thus, the added flexibility compromises for gaining the increased conversion rate (both enantiomers) with the loss in ee values. Hence, the aim should be to increase the rate of the reaction without increasing the flexibility of the biocatalyst; i.e., we are going to focus to have an improved K_{cat}/K_M for the favorable enantiomer by making the forward reaction faster for a particular process [5]. One of the ways is to use the vinyl esters as acyl donor.

3.1.5.1 Vinyl Esters: The Efficient Class of Donors

Among the esters routinely used for transesterification, vinyl esters have proved to be the most versatile acyl donors. Three esters, viz. vinyl acetate (I), vinyl propionate (II) and isopropenyl acetate (III), are mostly used (Fig. 3.1). For, both the product alcohol tautomerizes to a carbonyl compound: acetaldehyde in case of vinyl esters and acetone in case of isopropenyl acetate which make the forward reaction irreversibly fast [26].

The only disadvantage of vinyl esters is that some lipases (especially lipases from *Candida rugosa* and *Geotricum candidum*) cannot tolerate the liberated acetaldehyde which forms a Schiff base with a lysine residue near the active site [27] causing a loss in activity.

However, this deactivation was overcome to a limited extent by celite immobilization [27]. Alternatively, an expensive vinyl ester, viz. ethoxyvinyl acetate (IV, Fig. 3.1), has sometimes been used but has not gained enough popularity in the industries due to its high cost. Recent work shows that this deactivation followed by a decrease in E value could be overcome by co-cross-linking the lipase with bovine

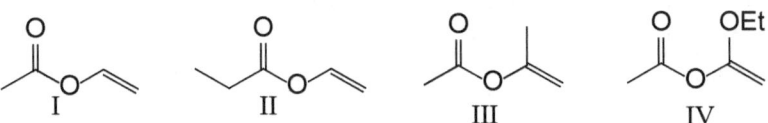

Fig. 3.1 Vinyl (or substituted vinyl) esters used as acyl donors

serum albumin [28]. Other acyl donors which are frequently used but with less applicability are acetoacetic ester, acetic anhydride and acetates of short chain primary alcohols.

3.1.5.2 Effect of Acyl Chain Length on E

As discussed in the previous section, vinyl esters have been the favorites for enzymatic transacylation reactions. Going ahead one step further, Ottoson and Hult have shown that even in vinyl esters a difference in the acyl chain length may result in different enantioselectivities (Scheme 3.1).

For this, the enzyme active site must be very sensitive toward the size of the acyl group. These researchers, while resolving racemic 3-methyl-2-butanol using a lipase from *Candida antarctica* (fraction B), have studied the effects of using vinyl propionate, vinyl butanoate, vinyl hexanoate and vinyl octanoate. Their report showed a steady increase in E from vinyl butanoate to vinyl octanoate, and an increase in E from 390 to 810 was observed [4]. Table 3.2 summarizes their findings.

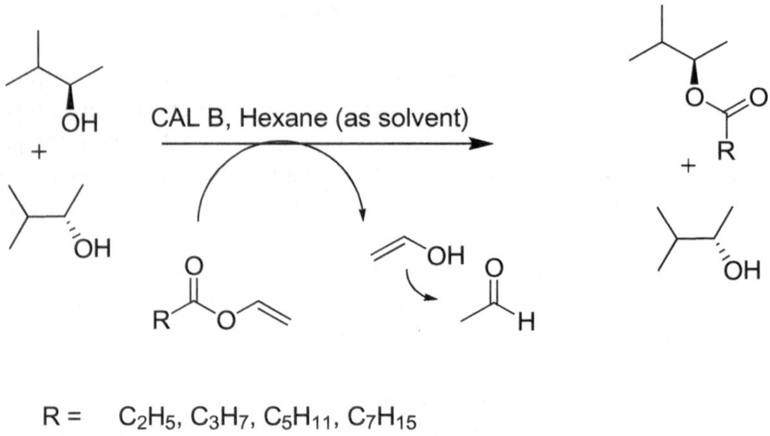

$$R = \quad C_2H_5, C_3H_7, C_5H_{11}, C_7H_{15}$$

Scheme 3.1 CALB catalyzed EKR of racemic 3-methyl-2-butanol using vinyl esters

Table 3.2 Change in E with the chain length of vinyl alkanoate taken as acyl donor for the kinetic resolution of racemic 3-methyl-2-butanol in hexane using CALB

Entry	Acyl donor	E (298 °K)	$\Delta G_{R,S}^{\ddagger}$ (kJ Mole^{-1})	$\Delta\Delta S_{R,S}^{\ddagger}$ (J Mole^{-1} °K^{-1})
1	Vinyl propanoate	470	− 15.3	− 11.5 ± 6.1
2	Vinyl butanoate	390	− 14.8	− 29.6 ± 6.0
3	Vinyl hexanoate	720	− 16.3	− 20.1 ± 4.5
4	Vinyl octanoate	810	− 16.6	− 25.9 ± 3.4

The increase in E (entry 2–4, Table 3.2) was caused by a significant decrease in Gibb's free energy of activation, $\Delta G^{\ddagger}$ and by a change in the entropy factor, $\Delta\Delta S^{\ddagger}$. Interestingly, it has been pointed out that these changes of thermodynamic parameters which were responsible for the change in E were more with the S-enantiomer. The lipase from *Candida antarctica* (fraction B) has got a deep narrow active site, like a "hair pin", and hence, the varying chain length of the vinyl ester (acyl donor) affects its activity. The size of the propanoate ester helps it to fit in a different way, and hence, an "out of trend" thermodynamic change happens causing an increased E.

3.1.5.3 High Enzyme-to-Substrate Concentration Ratio

In a continuation to the previous section, one of the other ways to make the process irreversible is to maintain a high enzyme-to-substrate concentration ratio, i.e., $\frac{[Enzyme]}{[Substrate]}$ should be $\gg 1$. Usually in biocatalytic reactions, this enzyme-to-substrate molar ratio is taken in the order of 1:100 [17]. Thus, this strategy may not sound much appealing for an industrial process for its lack in cost-effectiveness as most of the biocatalysts are not cheap. However, the idea of taking enzyme "as much as possible" seems to be very positive and may prove very effective for, if there are more active sites available, the possibility of the other (the misfit) enantiomer to get bound decreases. The cost-effectiveness, on the other hand, can be improved by using high performance biocatalyst formulations which can be recycled [26].

3.1.5.4 Lowering the Possibility of Diffusion Limitation

In many cases to improve the cost-effectiveness of the process, immobilized enzymes are used. The problem of diffusion is very common in carrier bound immobilized formulations, viz. in enzyme in a spherical particle or on a membrane. The ineffective assembly of the enzyme-carrier duo causes a hindrance to the substrate to move in or the product to move out.

Thus, there occurs an increased concentration of the substrate in the solution and a low substrate concentration within the support [17]. Both enantiomers would face the same difficulty to the same extent, but the effect would be more pronounced with the desired enantiomer. Thus, the rate of the reaction with the desired enantiomer would fall more rapidly as the enzyme naturally accepts it more in a usual situation resulting in a decrease in E value. In carrier free immobilization also, this is not very much uncommon.

In a unique work, while resolving racemic β-citronellol (Scheme 3.2) using cross-linked formulations of *B. cepacia* lipase, the extent of cross-linking was varied to study the change in E value. The molecule β-citronellol is a long chain primary alcohol with a remote chiral center. Resolving alcohols with remote chiral center has been difficult, and hence, an improved enantioselectivity was targeted by designing optimally flexible cross-linked enzyme aggregates (CLEA) for this substrate [5].

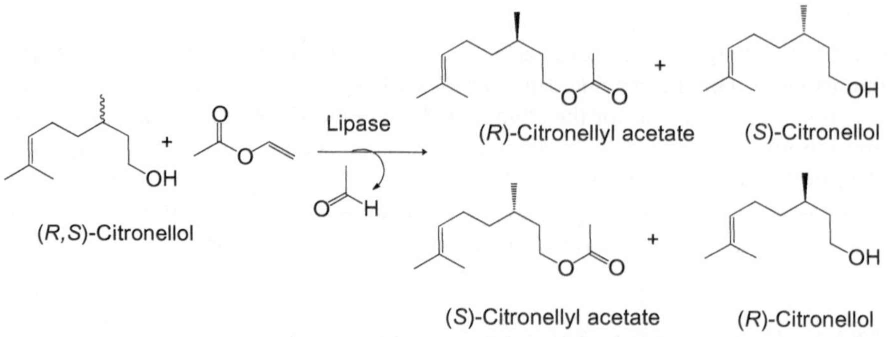

Scheme 3.2   EKR of racemic β-citronellol using vinyl acetate

Cross-linking was done using different concentrations of glutaraldehyde keeping the cross-linking time constant. Cross-linking makes enzyme less flexible and hence more enantioselective. Table 3.3 reflects these results.

The downfall in E from 74 to 6 was found to be due to diffusion limitation caused by the over cross-linking effected by a 6× concentrated cross-linking solution of glutaraldehyde resulting in a rock like formulation constituting of large particles which was found by SEM analysis [5].

Table 3.3   Variation of E with the degree of cross-linking in *B. Cepacia* Lipase (BCL) in an enantioselective transacylation of racemic β-citronellol

BCL forms	Glutaraldehyde (mM)	Particle (solid)	E	V_{max} m mol mg^{-1} h^{-1}	K_M m mol	V_{max}/K_M mg^{-1} h^{-1}	SEM analysis
CLEA A	10	< 5 µm	74	6.51	3.87	1.7	
CLEA B	40	5–10 µm	11	4.02	5.72	0.7	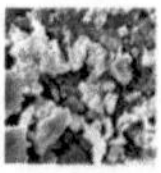
CLEA C	60	Large clusters	6	2.45	7.47	0.3	

Adapted with permission from [5]. Copyright (2008), Taylor and Francis UK Ltd.

3.2   Uses of Lipases for Kinetic Resolution: Examples

Table 3.4 shows a wide range of substrate including primary alcohols [21–24, 29–32], secondary acyclic alcohols [25, 26, 33, 34] and cyclic (bicyclic and tricyclic) alcohols [27, 28, 35, 36]. Out of these, resolving primary alcohols are the most challenging ones as the reactive end is not close to the chiral center [5, 15, 21–23, 29–31].

These could be resolved by enantioselective transacetylation catalyzed by lipases taking vinyl esters as acyl donor. Lipase catalyzed synthesis of some very important chiral pharmaceutical intermediates has been short-listed in Table 3.5. In all the cases, an *ee* > 95% with a conversion > 40% was reported with a fair enantioselectivity [29–35, 37–43] A range of drugs or drug intermediates including antihypertensive drugs [29, 30, 37, 38], anti-cholesterol drugs [31, 39], anti-inflammatory drugs [34, 35, 42, 43], anti-convulsant drugs [33, 41] and anticancer drugs (chemotherapeutic agent) was synthesized [29, 37] by lipase catalyzed enantioselective biotransformations (Table 3.5).

3.3   Use of Proteases for EKR

Proteases, as well, find application in the kinetic resolution of secondary alcohols [44], a wide variety of racemic amino acids [45], and in the regioselective transacetylation of sugar derivatives [46]. Among all the proteases, the most popular for stereoselective biotransformation is the subtilisin Carlsberg protease (from *Bacillus licheniformis*) [47]. Table 3.6 illustrates the use of the proteases in organic synthesis. Among the wide range of applications, resolution to chirally pure non-natural amino acids is increasingly getting more attention due to their extensive use in the synthesis of peptidic drugs [48]. A series of N-protected (or free) amino acids were resolved by enantioselective hydrolysis using protease subtilisin Carlsberg of which hydrophobic amino acids, viz. *tert*-leucine (entry 2, Table 3.6), due to its bulky side chain used increasingly as a building block for the synthesis of chiral auxiliaries. Alcalase® is a commercially available formulation of subtilisin. The fourth type of reaction (entry 4, Table 3.6), however, is a regioselective esterification of a sugar derivative.

Though subtilisin seems to catalyze similar type of reactions (with esters) as lipases, it differs widely in terms of substrate specificity. The next section deals with the basic differences in the stereoselective biotransformations catalyzed by these two hydrolases.

Table 3.4 Uses of lipases in kinetic resolution of racemic alcohols using vinyl esters as acyl donors

Entry	Systems	Resolved enantiomer	Hydrolase	References
1			P. fluorescence lipase	[29]
2			Lipase AK	[30]
3			B. cepacia lipase	[31]
4			Porcine pancreatic lipase	[32]
5			P. fluorescence lipase	[33]
6			B. cepacia lipase	[34]
7			Lipase AK	[35]
8	(+,-)		B. cepacia lipase	[36]

Table 3.5 Uses of lipases in synthesis of chiral drug intermediates

Entry	Drugs	Drug or drug precursor (synthesized enzymatically)	Hydrolase	References
1	Captopril, enalapril, lisinopril (antihypertensive drugs)		*B. cepacia* lipase (Lipase PS 30)	[37]
2	Monopril (antihypertensive drug)		PPL and lipase from *C. viscosum*	[38]
3	Taxol (chemotherapeutic agent)		*B. cepacia* lipase, BMS lipase (extracellular lipase from fermented *Pseudomonas sp.* SC13856)	[37]
4	Pravastatin (anti-cholesterol drug)		Pig liver esterase	[39]
5	Propranolol (beta adrenergic blocking agent)		*B. cepacia* lipase	[40]
6	Thioridazine (antipsychotic) Piperadol (anti-convulsant)	*R*-(+)-pipecolic acid	Lipase from *Asp. niger*	[41]

(continued)

Table 3.5 (continued)

Entry	Drugs	Drug or drug precursor (synthesized enzymatically)	Hydrolase	References
7	Ibuprofen (anti-inflammatory drug)		Novozym 435 (CALB)	[42]
8	Naproxen (anti-inflammatory drug)		*C. rugosa* lipase	[43]

Table 3.6 Uses of proteases in stereoselective synthesis

Entry	Systems	Resolved enantiomer	Hydrolase	References
1			Subtilisin	[45]
2			Subtilisin	[48]
3			Alcalase	[47]
4			Alcalase	[46]

3.4   Opposite Enantiopreference by Hydrolases: The Kazlauskas' Model

During a kinetic resolution, which enantiomer will be preferred by a hydrolase (at least by a lipase or a protease) can be predicted following Kazlauskas model [21, 44]. The readers must be careful in dealing with these models for these are valid for only secondary alcohol and a primary amine.

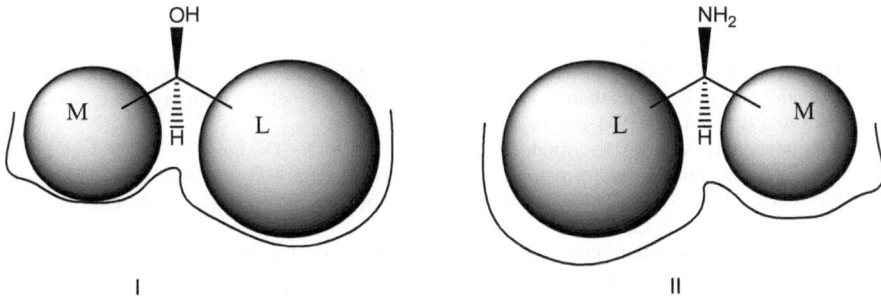

Fig. 3.2 Empirical rule based on Kazlauskas's model for enantiopreference of lipases toward secondary alcohols and of protease subtilisin for secondary alcohols or chiral amines

It is assumed that the priority order is: OH or substituted OH > Large (L) group > Medium (M) group > H, the smallest group. Lipases and subtilisin protease show opposite enantiopreference when the functional group is directly attached (non-remote) to the reaction site.

The rule is based upon the size of the substituents at the chiral center. High enantioselectivity is observed when one substituent is of a medium size (e.g., methyl group) and the other substituent is large (e.g., aromatic ring).

According to the empirical rules based upon the models shown in Fig. 3.2:

- Model I is applied for lipases in a kinetic resolution of secondary alcohols; (*R*)-enantiomer would be preferred over the (*S*)-enantiomer. The model proved valid for a vast majority of compounds in reactions catalyzed by lipases from *Burkholderia cepacia* and *Candida rugosa* [14].
- Model II is applied for proteases (especially, subtilisin) which shows an active site feature which is a mirror image of the Model I. Thus for chiral amines and analogous secondary alcohols, (*S*)-enantiomer is preferred by subtilisin (or a protease or esterase, e.g., pig liver esterase with identical active site features).
- The models are applicable only when the functional group (OH or NH_2) is directly attached to the chiral center. When the chiral center is remote, i.e., flanked by even a single methylene group, the enantiopreference becomes unpredictable.

References

1. Ema T, Yamaguchi K, Wakasa Y, Yabe A, Okada R, Fukumoto M, Yano F, Korenaga T, Utaka M, Sakai T (2003) Transition-state models are useful for versatile biocatalysts: kinetics and thermodynamics of enantioselective acylations of secondary alcohols catalyzed by lipase and subtilisin. J Mol Catal B Enz 22:181–192
2. Persson M, Costes D, Wehtje E, Adlercreutz P (2002) Effects of solvent, water activity and temperature on lipase and hydroxynitrile lyase enantioselectivity. Enzyme Microb Technol 30:916–923

3. Sakai T (2004) Low-temperature method'for a dramatic improvement in enantioselectivity in lipase-catalyzed reactions. Tetrahedron Asymm 15:2749–2756

4. Ottoson J, Hult K (2001) Influence of acyl chain length on the enantioselectivity of *Candida antarctica* lipase B and its thermodynamic components in kinetic resolution of sec-alcohols J Mol Catal B Enz 11:1025–1028

5. Majumder AB, Mondal K, Singh TP, Gupta MN (2008) Designing cross-linked lipase aggregates for optimum performance as biocatalysts. Biocatal Biotransform 26(3):235–242

6. Bornscheuer UT (2002) Methods to increase enantioselectivity of lipases and esterases. Curr Opin Biotechnol 13:543–547

7. Wehtje E, Costes D, Adlercreutz P (1997) Enantioselectivity of lipases: effects of water activity. J Mol Catal B Enzym 3(5):221–230

8. Homberg E, Hult K (1990) Dependence of the enantiomeric ratio and the reaction rate on the proportions of water and cyclohexane. Biocatal Biotransform 3:243–251

9. Reslow M, Adlercreutz P, Mattiasson B (1992) Modification of the microenvironment of enzymes in organic solvents. Substitution of water by polar solvents. Biocatal Biotranform 6:307–318

10. Wehtje E, Costes D, Adlercreutz P (1997) Enantioselectivity of lipases: effects of water activity. J Mol Catal B Enzym 3:221–230

11. Nordin O, Hedenström E, Högberg H-E (1994) Enantioselective transesterifications of 2-methyl-1-alcohols catalysed by lipases from *Pseudomonas*. Tetrahedron Asymm 5:785–788

12. Orrenius C, Norin T, Hult K, Carrea G (1995) The *Candida antarctica* lipase B catalysed kinetic resolution of seudenol in non-aqueous media of controlled water activity. Tetrahedron Asymm 6:3023–3030

13. Hogberg HE, Edlund H, Berglund P, Hedenstrom, H (1993) Water activity influences enantios-electlvlty in a lipase-catalysed resolution by esterificatlon in an organic solvent. Tetrahedron Assym 4:2123–2126

14. Bornscheuer UT, Herar A, Kreye L, Wendel V, Capewell A, Meyer HH, Scheper T, Kolisis FN (1993) Factors affecting the lipase catalyzed transesterification reactions of 3-hydroxy esters in organic solvents. Tetrahedron Asymm 4:1007–1016

15. Majumder AB, Shah S, Gupta MN (2007) Enantioselective transacetylation of (R,S)-beta-citronellol by propanol rinsed immobilized Rhizomucor miehei lipase. Chem Cent J (BMC Chem) 1:10

16. Wu S-H, Chu F-Y, Wang K-T (1991) Reversible enantioselectivity of enzymatic reactions by media. Bioorg Med Chem Lett 1:339–342

17. Straathof AJJ, Jongejan JA (1997) The enantiomeric ratio: origin, determination and prediction. Enzyme Microbe Technol 21:559–571

18. Secundo F, Riva S, Carrea G (1992) Effects of medium and of reaction conditions on the enantioselectivity of lipases in organic solvents and possible rationales. Tetrahedron Asymm 3(2):267–328

19. Gerlach D, Schreier P (2009) Esterification in organic media for preparation of optically active secondary alcohols: effects of reaction conditions. Biocatal Biotransform 2(4):257–263

20. Sakurai T, Margolin AL, Russell AJ, Klibanov AM (1988) Control of enzyme enantioselectivity by the reaction medium. J Am Chem Soc 110(21):7236–7237

21. Tawaki S, Klibanov AM (1992) Inversion of enzyme enantioselectivity mediated by the solvent. J Am Chem Soc 114:1882–1884

22. Patel RN (2000) Stereoselective biocatalysis. Mercel –Dekker, New York

23. Sheldon RA, Schoevaart R, Van Langen LM (2005) Cross-linked enzyme aggregates (CLEAs): a novel and versatile method for enzyme immobilization (a review). Biocatal Biotransform 23:141–147

24. Yu HW, Chen B, Wang X, Yang YY, Ching CB (2006) Cross- linked enzyme aggregates (CLEAs) with controlled particles: application to Candida rugosa lipase. J Mol Catal B Enzym 43(1–4):124–127

25. Mateo C, Palomo JM, van Langen LM, van Rantwijik F, Sheldon RA (2004) A new, mild cross-linking methodology to prepare cross-linked enzyme aggregates. Biotechnol Bioeng 86:273–276

26. Gupta MN (ed) (2000) Methods in non aqueous enzymology. Birkhauser-Verlag, Basel, Switzerland ·
27. Weber HK, Zuegg J, Faber K, Pleiss J (1997) Molecular reasons for lipase-sensitivity against acetaldehyde. J Mol Catal B Enz 3:131–138
28. Majumder AB, Gupta MN (2010) Stabilization of *Candida rugosa* lipase during transacetylation with vinyl acetate. Bioresource Technol 101:2877–2879
29. Baczko K, Larpent C (2000) Lipase-catalyzed transesterification of primary alcohols: resolution of 2-ethylhexan-1-ol and 2-ethylhex-5-en-1-ol. J Chem Soc. Perkin Trans 2:521–526
30. Berkowitz DB, Pumphrey JA, Shen Q (1994) Enantiomerically enriched α-vinyl amino acids via lipase-mediated "reverse transesterification". Tetrahedron Lett 35:8743–8746
31. Miyaoka H, Sagawa S, Inoue T, Nagaoka H, Yamada Y (1994) Efficient synthesis of optically active cyclohexenones. Chem Pharm Bull 42:405–407
32. Van Tol JBA, Kraayveld DE, Jongejan JA, Duine JA (1995) The catalytic performance of pig pancreas lipase in enantioselective transesterification in organic solvents. Biocatal Biotransform 12:119–136
33. Brown SM, Davies SG, de Sousa JAA (1993) Kinetic resolution strategies II: enhanced enantiomeric excesses and yields for the faster reacting enantiomer in lipase mediated kinetic resolutions. Tetrahedron Asymm 4:813–822
34. Nakamura et al (1995) Enantiomer separation (Toda F (ed)). Springer, Netherlands
35. Iimori T, Azumaya I, Hayashi Y, Ikegami S (1997) A practical preparation of optically active endo-bicyclo [3.3. 0] octen-2-ols. Chem Pharm Bull 45:207–208
36. Jan Willem JF, Thuring GHL, Margreth AN, Antonius W, Klunder JH, Zwanenburg B (1996) Enzymatic kinetic resolution of 5-Hydroxy-4-oxa-*endo*-tricyclo[5.2.1.0^{2,6}]dec-8-en-3-ones: a useful approach to D-Ring synthons for strigol analogues with remarkable stereoselectivity. J Org Chem 61:6931–6935
37. Patel RN (1997) Stereoselective biotransformations in synthesis of some pharmaceutical intermediate. Adv Appl Microbiol 43:91–140
38. Patel RN, Robison RS, Szarka J (1990) Stereoselective enzymatic hydrolysis of 2-cyclohexyl- and 2-phenyl-1,3-propanediol diacetate in biphasic systems. Appl Microbiol Biotechnol 34:10–14
39. Suemune H, Takahashi M, Maeda S, Xie ZF, Sakai K (1990) Asymmetric hydrolysis of cis, cis-5-benzyloxy-1,3-diacetoxy-cyclohexane and its application to the synthesis of chiral lactone moiety in compactin. Tetrahedron Asymm 1(7):425–428
40. Borowiecki P, ZdunB, Popow N, Wiklińska M, Reiter T, Kroutil W (2022) Development of a novel chemoenzymatic route to enantiomerically enriched β-adrenolytic agents. A case study toward propranolol, alprenolol, pindolol, carazolol, moprolol, and metoprolol. RSC Adv 12:22150–22160
41. Christine M, Chen NY, Serreqi AN, Huang Q, Kazlauskas RJ (1994) Kinetic resolution of pipecolic acid using partially-purified lipase from *Aspergillus niger*. J Org Chem 59:2075–2081
42. Trani M, Ducret A, Pepin P, Lortie R (1995) Scale-up of the enantioselective reaction of the enzymatic resolution of (*R,S*)-ibuprofen. Biotechnol Lett 17:1095–98
43. Gu QM, Chen CS, Sih CJ (1986) A facile enzymatic resolution process for the preparation of (+)-S-2-(6-hethoxy-2-naphthyl) propionic acid (Naproxen). Tetrahedron Lett 27:1763–1766
44. Kazlauskas RJ, Weissfloch ANE (1997) A structure-based rationalization of the enantiopreference of subtilisin toward secondary alcohols and isosteric primary amines. J Mol Catal B Enz 3:65–72
45. Miyazawa T (1999) Enzymatic resolution of amino acids via ester hydrolysis. Amino Acids 16:191–213
46. Riva S, Chopineau J, Kieboom APG, Klibanov AM (1988) Protease-catalyzed regioselective esterification of sugars and related compounds in anhydrous dimethylformamide. J Am Chem Soc 110(2):584–589

47. Chen ST, Tu CC (1993) War KT (1993) Selective incorporation of d-amino acid esters into peptides catalyzed by alcalase in t-butanol. Bioorg Med Chem Lett 3(4):539–542
48. Agosta E, Caligiuri A, D'Arrigo P, Servi S, Tessaro D, Canevotti R (2006) Enzymatic approach to both enantiomers of N-Boc hydrophobic amino acids. Tetrahedron Asymm 17(13) 14:1995–1999

Chapter 4
Dynamic Kinetic Resolution

Abstract In stereoselective biocatalysis, kinetic resolution is replaced wherever possible by dynamic kinetic resolution to have a better yield of the enzyme chosen enantiomeric product. Many transitional metal complexes have been found to reracemize the enantiomeric substrate (relatively untouched by the enzyme) under the milder conditions (of enzymatic kinetic resolution) and thereby avoiding the harsh conditions aided by chemicals. This improves the atom economy and prevents the unwanted side reactions during racemization. In this chapter, some of the very widely used and effective strategies with examples have been put forward.

Keywords Dynamic kinetic resolution · Reracemization · Shvo's catalysts · Reverse enantiopreference · Hydrolases with nanopalladium

4.1 DKR: Understanding the Process

As mentioned earlier in chapters, while dealing with the demerits of EKR, extensive research has been focused to get a proper utilization of the enzyme unfriendly enantiomer to have a better yield and atom efficiency. One of the processes of achieving this is "reracemization" [1, 2]. In this process, the relatively untouched enantiomer (50% of any perfect racemic mixture of the starting compound) is converted to a racemic mixture of which again 50%, as is picked up by the enzyme, gets converted to the desired product enantiomer. This process is repeated, and each cycle results in an increase in the product concentration.

Let us try to understand this with an illustrative model. If a reaction starts with a racemic mixture with an initial concentration X mmol, after a perfect EKR, it leaves 0.5X mmol of the unused enantiomer. After reracemization, it gives 0.5X mmol of racemic mixture (assuming 100% racemization). Consequently, after EKR of this newly racemized fraction (assuming an ideal situation) 0.25X mmol would be converted to product in the second cycle. A repeat of this process would yield 0.125X mmol of product, and thus, the concentration of the relatively untouched enantiomer decreases after each repetition of "reracemization followed by EKR" exhibiting

A. B. Majumder and K. V. S. Ranganath, *Understanding Kinetic Resolution by Hydrolases*, SpringerBriefs in Molecular Science, https://doi.org/10.1007/978-3-031-46353-2_4

0.25X mmol, 0.125X mmol, 0.0625X mmol, 0.0315X mmol, 0.01575X mmol, 0.007875X mmol and 0.0039375X mmol, and thus, it goes on till no substrate is left. Likewise after each cycle, the product concentration increases by 0.5X, 0.5X + 0.25X, 0.5X + 0.25X + 0.125X, 0.5X + 0.25X + 0.125X + 0.0625X and so on. The process is thus continued till complete, and no racemic mixture is left.

In most of the cases, chemical racemization techniques required harsh reaction conditions such as application of heat, uses of acids or bases and went through chirally unstable intermediates. As a consequence of the harsh reaction conditions, the possibility of undesired side reactions, such as elimination, condensation, rearrangement, and/or decomposition increase causing a low yield and of less preparative utility with a poor atom economy. Thus, efforts were made to make racemization in milder condition, of more synthetic utility and, friendly to the enzymes (in terms of activity and reusability), which gradually paved the way to the use of transition metal complexes for this purpose [3]. Very good examples involve the uses of ruthenium and palladium for the racemization of the relatively untouched substrate enantiomer in combination with biocatalysts. In the following sections, we will be focusing some of very interesting examples of uses of organometallics in DKR aided by hydrolases.

4.2 DKR of Secondary Alcohols and Amines: Use of Ruthenium

4.2.1 Use of Lipase-Ruthenium Combination

One mostly used organometallic reracemizer is a dimer of dicarbonyl tetraphenyl hydroxy cyclopentadienyl ruthenium complex, the Shvo's catalyst. It is a white crystalline solid which is stable at room temperature in air. The catalyst is prepared by refluxing ruthenium carbonyl complex with tetraphenylcyclopentadienone in the presence of methanol at 40 °C (Scheme 4.1). It is a metal–ligand bifunctional catalyst wherein redox activity is distributed between metal center and a ligand. This catalyst has been extensively used in the dehydrogenation of alcohols in the presence or absence of hydrogen acceptors.

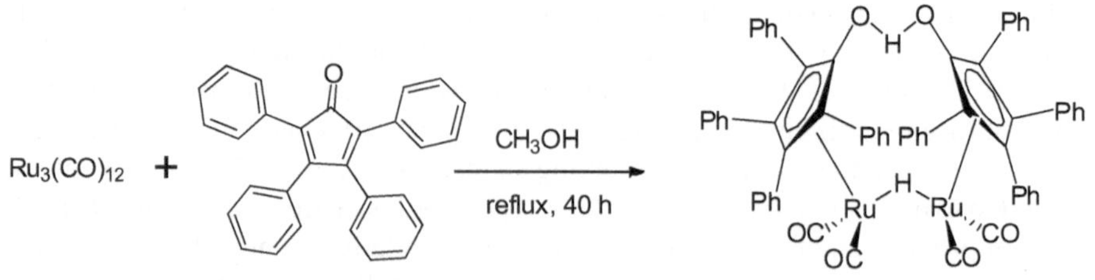

Scheme 4.1 Shvo's catalyst preparation

This ability of abstracting hydrides makes it an excellent catalyst to convert secondary alcohols to a ketone and help in reracemization (as the process of hydride shift is a reversible one) in dynamic kinetic resolution. Scheme 4.2 shows this process. The catalyst is in rapid equilibrium with its zwitterionic form. This monomeric form through the positively charged ruthenium takes up hydride from the one enantiomer, the one relatively untouched by enzyme (shown in blue), of alcohol and converts it

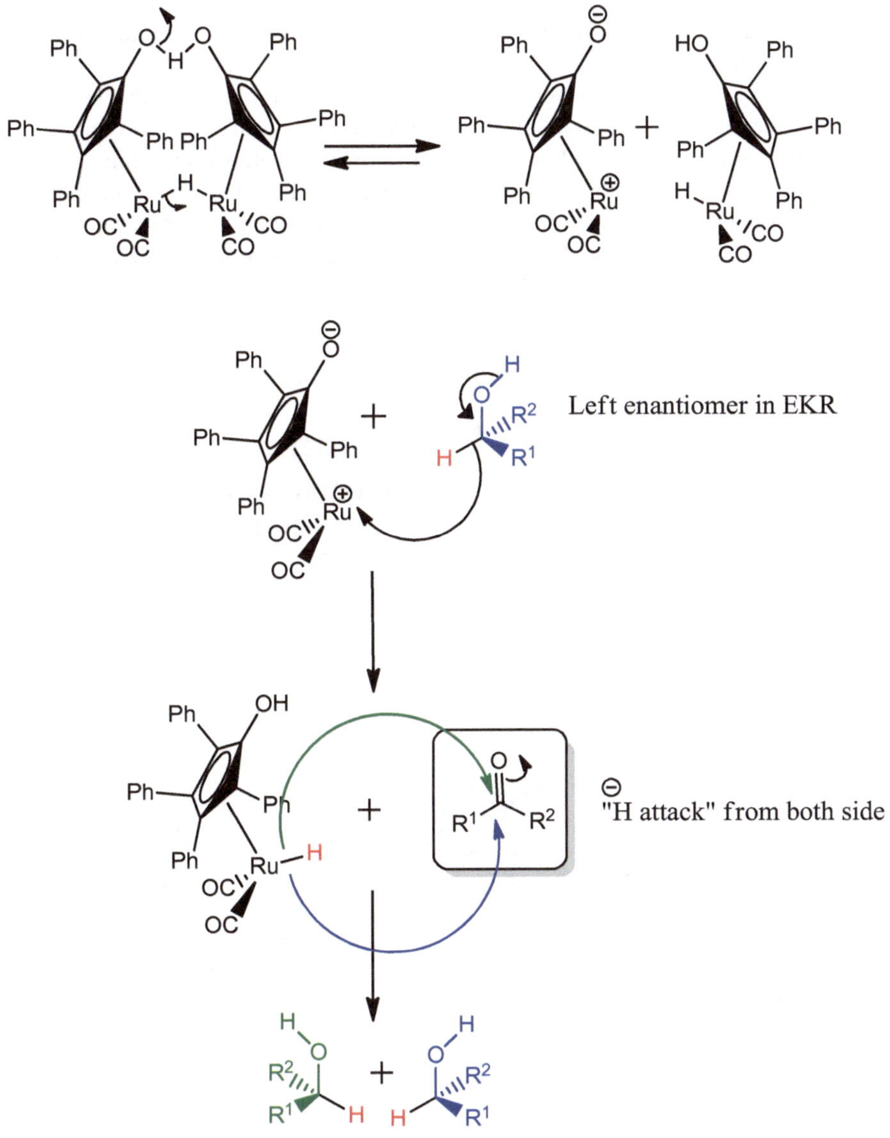

Scheme 4.2 Reracemization of a secondary chiral alcohol

into a flat ketone. The carbonyl carbon of this ketone then undergoes the attack of the hydride from the same ruthenium from both above and below the plane and thereby forms the racemic alcohol. This newly formed racemic mixture then is utilized in enzymatic kinetic resolution for the synthesis of enantiopure acylated alcohols.

For instance, DKR of various secondary alcohols has been carried out in the presence of Novozyme 435 which is a commercially available immobilized formulation of *Candida antarctica* lipase B using 4-chlorophenyl acetate as the acyl donor with the help of Shvo's catalyst. Both aliphatic and aromatic secondary alcohols responded well under the reaction conditions and gave the products in > 99% *ee* with up to 88% yields (Scheme 4.3).

One major problem of using Shvo's catalyst when combined with CALB is a slow reaction and/or reracemization which may take 7 days in cases to complete. To address this issue, a chlororuthenium analog of Shvo's catalyst (Scheme 4.4) was used. With this, similar substrates (1-phenylethanol analogs) could be reracemized within 10 min of activation by potassium *tert*-butoxide, and the chiral acetate could be obtained in 99% yield with 99% *ee* within 4 h when isopropenyl acetate was used as the acyl donor (Scheme 4.4) [4].

Scheme 4.3 Dynamic kinetic resolution in the presence of enzymes with Shvo's catalyst

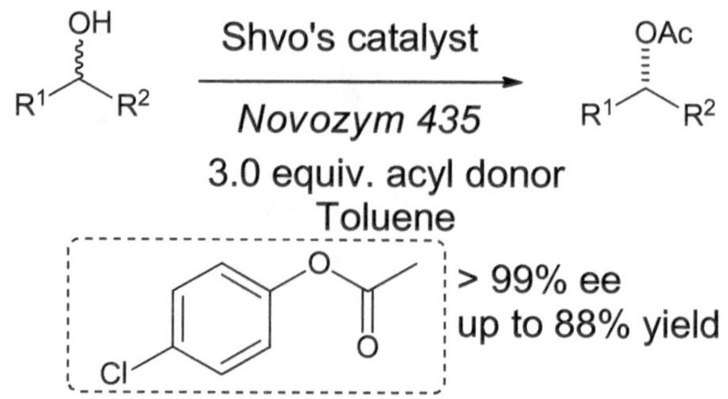

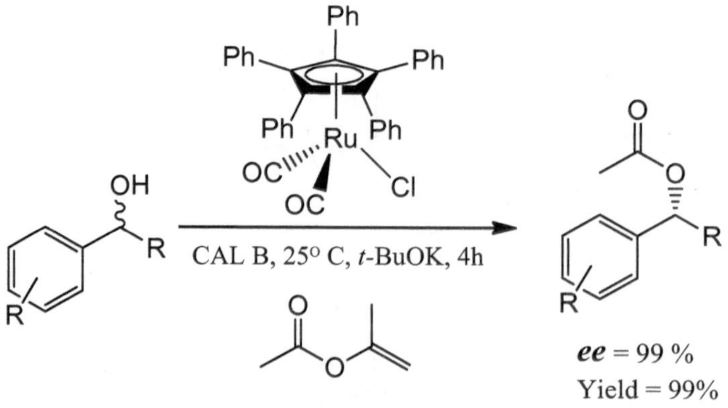

Scheme 4.4 Shvo's catalyst: use of chlororuthenium system

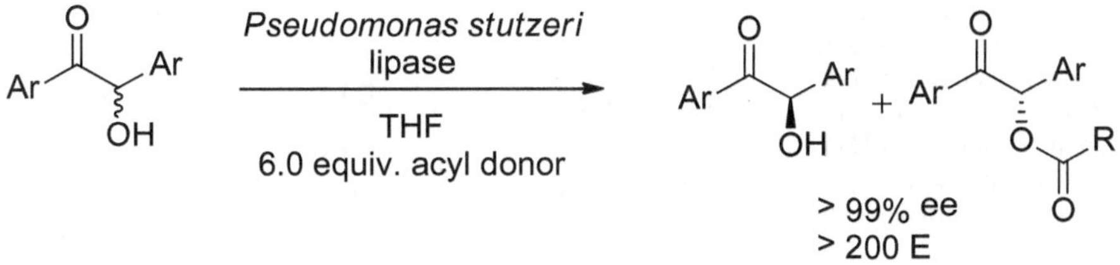

Scheme 4.5 Dynamic kinetic resolution of benzoins

Use of ruthenium has been found fruitful also in benzoins. These α-hydroxy aromatic ketones are useful building blocks for the synthesis of heterocyclic compounds and also useful as urease inhibitor. The DKR of benzoins was reported using lipase in combination with ruthenium catalyzed substrate racemization (Scheme 4.5). The acylated products with high *ee* were isolated in excellent yields using an excess (up to 6 times of substrate concentration) of acyl donors. The results of DKR with *Pseudomonas stutzeri* lipase in combination with ruthenium complex was remarkable as instead of preferring (*R*) it yielded (*S*)-acetate with benzoins which violated Kazlauskas' rule. It has been pointed out that the benzoyl group is forced to occupy the alternate pocket as the carbonyl group is bound to the active site of CALB [5].

Interestingly, a similar type of effect of (*S*)-enantiopreference could be observed during the DKR of α-hydroxyesters with *Pseudomonas cepacia* (also known as *Burkholderia cepacia*) lipase in the presence of Shvo's catalyst (Scheme 4.6) [7]. Once again, perhaps, the binding interaction of carboxylic carbon with enzyme plays a role to alter the enantiopreference (Sect. 3.4, Fig. 4.1).

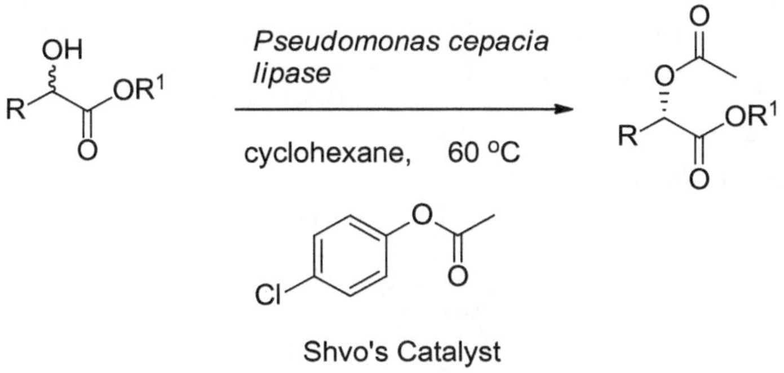

Scheme 4.6 Resolution of α-hydroxyester using Shvo's catalyst and *Pseudomonas cepacia* lipase

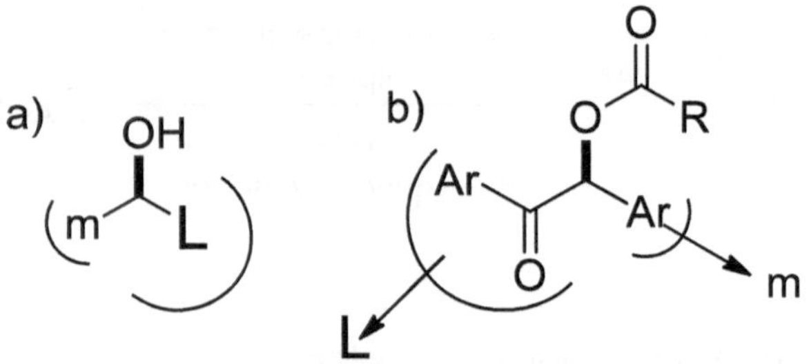

Fig. 4.1 Ruthenium–lipase combo reactions of benzoins: reverse enantiopreference. Adapted with permission from [6]. Copyright 1997, Elsevier Science & Technology Journals

4.2.2 DKR: Use of Subtilisin–Ruthenium Combination

For secondary alcohols, as we have seen before, lipases (unless genetically engineered or otherwise modified suitably) usually prefer R-enantiomer, and these are of less use where getting the S-enantiomer of secondary alcohol is more important. One alternative is the use of a protease from subtilisin Carlsberg. Subtilisin with a complementary active site feature (shown in Chap. 3, Sect. 3.4) prefers the S-alcohol. However, in non-aqueous media the commercial enzyme formulation (taken straight from the bottle) has got stability issues and is found to exhibit low activity. To overcome this, Kim et al. used surfactant-treated subtilisin (Scheme 4.7).

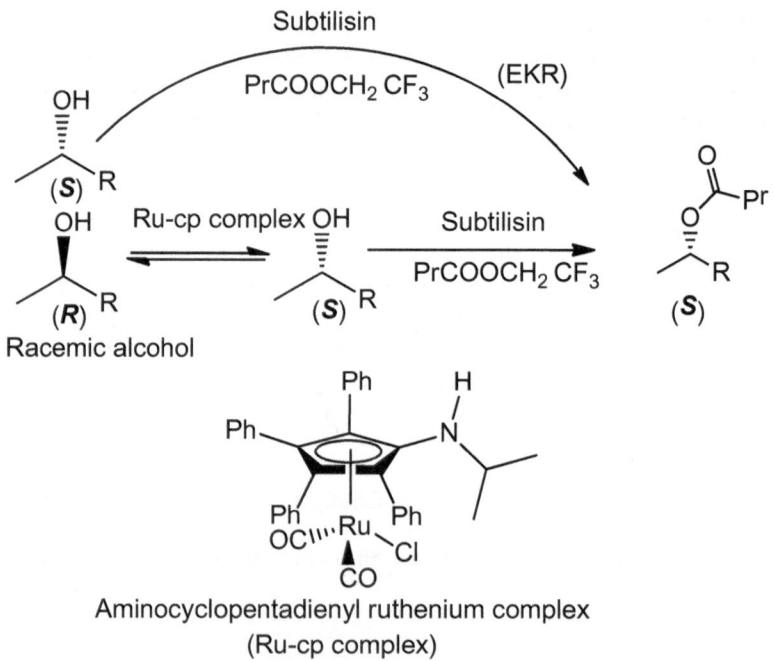

Scheme 4.7 Use of surfactant-treated subtilisin with ruthenium

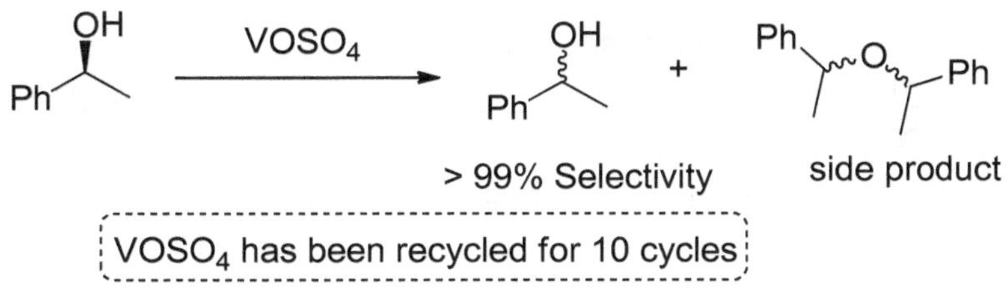

> 99% Selectivity   side product

VOSO$_4$ has been recycled for 10 cycles

Scheme 4.8 Vanadium sulfate: racemization selectivity and reusability

With trifluoroethyl butanoate as an acylating agent and an enzyme compatible tetraphenylcyclopentadienylamino dicarbonyl ruthenium chloride, they achieved the DKR that yielded (*S*)-products in good yields with high optical purities at room temperature [8].

Similar results were achieved in ionic liquids using cymene-ruthenium complex. The use of ionic liquid made the biocatalyst formulation reusable many times [9]. In this connection, it is worth mentioning that VOSO$_4$ has been found to be remarkably retaining its activity while reusing. In a racemization reaction of (*S*)-1-phenyl ethanol at 50 °C, VOSO$_4$ has been utilized as a heterogeneous catalyst (Scheme 4.8). The recyclability of VOSO$_4$ in the racemization was studied with one hour reaction time and was found to be active for ten cycles. However, in this process, each cycle of reracemization gives a side product.

4.3 DKR of Secondary Alcohols and Amines: Use of Palladium

4.3.1 Palladium as Heterogeneous Catalyst

Pd as heterogeneous catalyst is best used for the DKR of secondary amines. The first amine deracemization was tried using ethyl acetate as the acyl donor with Pd/C catalyst in combination with CALB. The reaction took nearly a week. Even starting with oximes could not improve the reaction rate. Lately, Pd was found to be well fitted in with a combination of alkaline earth metal salts. Thus when used with barium sulfate with CALB, racemic 1-aryl ethylamines exhibited 99% product *ee* during the synthesis of(*R*)-amides (Scheme 4.9) with a very good yield of 90% [10].

The mechanism of racemization involves the formation of the ketone from the secondary alcohol by the reductive elimination of hydrogen to the palladium(II) from the carbonyl carbon (Scheme 4.10) [11]. Likewise, in amine, the racemization happens through the formation of imine.

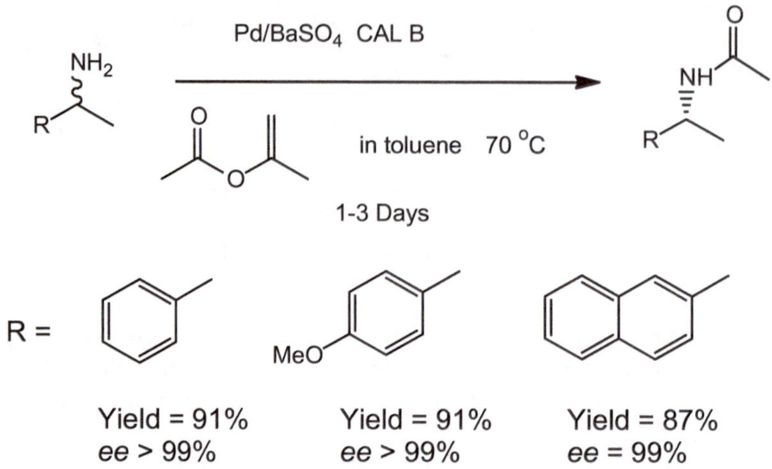

Scheme 4.9 Pd-CALB catalyzed DKR of benzylic amines

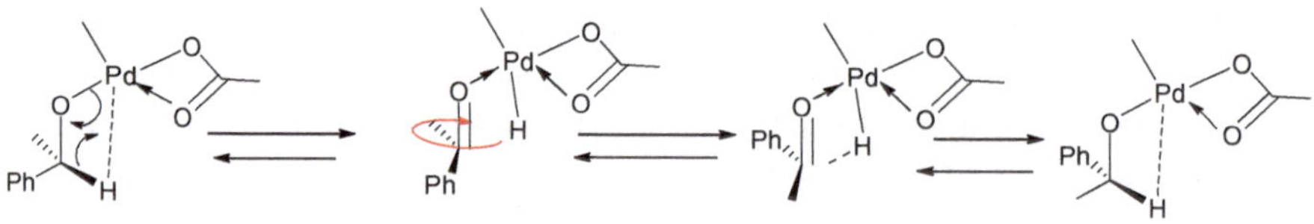

Scheme 4.10 Reracemization by Pd(II) catalyst for secondary alcohol

Palladium has been successfully used for the enantioselective hydrolysis of allyl acetates (Scheme 4.11) in combination with CALB in a buffer solution. In this case, two organometallic compounds, viz. Pd(PPh$_3$)$_4$ and 1,1′-bis(diphenylphosphino)ferrocene (dppf), were used as a combo "racemizing catalyst system". However, the process was found to be with limited scope for it failed to give consistent results with cyclic acetates, viz. cyclohexenyl acetates. However, palladium has been found to be an excellent catalyst for the racemization of enantiomeric amines [12].

4.3.2 Use of Palladium Nanocatalyst

The strategy was also successfully applied in chiral amines explored with nanochemistry. Thus, the dynamic kinetic resolution of 1-phenylethylamine was carried out in combination of Pd nanoparticles/AlO(OH) and lipase (Novozym 435) aided transacylation in toluene using methoxyethyl acetates as acyl donors (Scheme 4.12). The

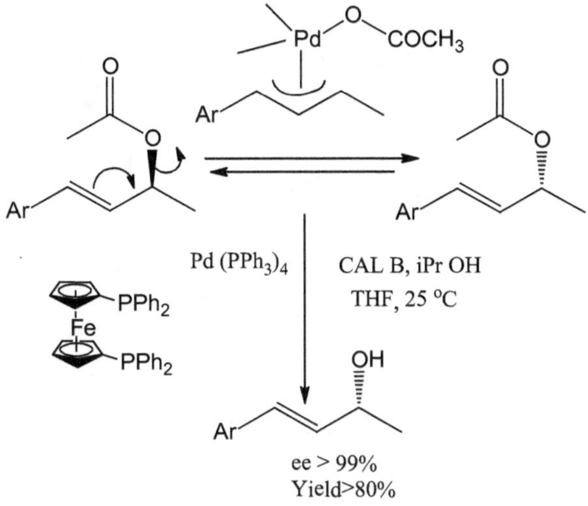

Scheme 4.11 Racemization of allyl acetates: use of palladium

corresponding products (R)-N-acyl-1-phenylethyl amines were isolated with high optical purities and in good yields. A considerable decrease in enantiomeric excess of the product has been observed as the phenyl group from the chiral center moves to β-position [13].

A combination of *Candida antarctica* lipase A (CALA) and the nanopalladium catalyst has been found to be excellent for dynamic kinetic resolution of β-amino esters (Scheme 4.13).

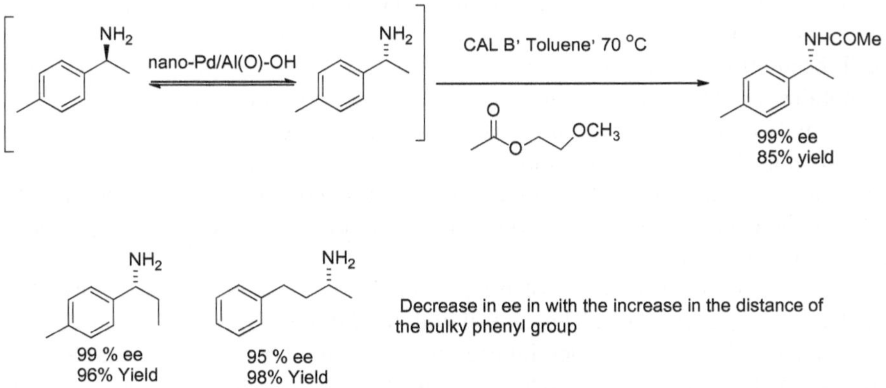

Scheme 4.12 Dynamic kinetic resolution using Pd nanoparticles

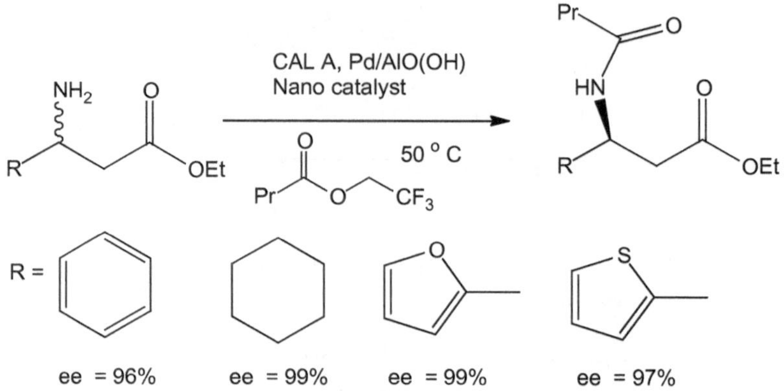

Scheme 4.13 Use of CALA in combination with palladium nanocatalyst for the resolution of beta amino esters

The process was found to be very much fruitful when the beta alkyl group is cyclic: worked well for both aromatic and aliphatic six-membered rings and five-membered aromatic heterocycles. In all the cases, the yield was excellent (> 99%) with 96–99% *ee* of (*S*)-amides [14].

4.3.3   Use of Pd-Enzyme Immobilized Hybrid Catalyst

Siliceous mesocellular foams (MCF) are mesoporous silica materials with large surface area and high loading capacity of enzymes as immobilization material. It contains high concentration of surface silanol groups that help in tagging with many functional groups. Using the strategies earlier applied in CALA immobilized on MCF been utilized with Shvo's catalyst or Pd nanocatalyst, Pd/Al O(OH) [15] co-immobilized CALB and the catalyst on MCF support and thereby ensuring each MCF cavity contains a lipase unit and a nano-Pd. The idea was to create a compact hybrid catalyst in a porous structure which acted like an artificial metalloenzyme. Pd nanoparticles were immobilized on aminopropyl functionalized MCF, and then the enzyme molecules were immobilized via cross-linking with the aminopropyl MCF using glutaraldehyde.

This hybrid catalyst could resolve 1-phenylethylamine magically with 99% yield and 99% *ee* (Scheme 4.14). The control with the Pd immobilized MCF and CALB immobilized MCF proved much inferior and gave only 66–89% yield with 99% *ee* of (*R*)-*N*-methoxyacetyl-1-phenylethylamine [15]. This novel work showed that the enzyme and nano-Pd catalyst in close proximity as a unit on a porous matrix is a much better option as a combo rather than simply a mixture of the two.

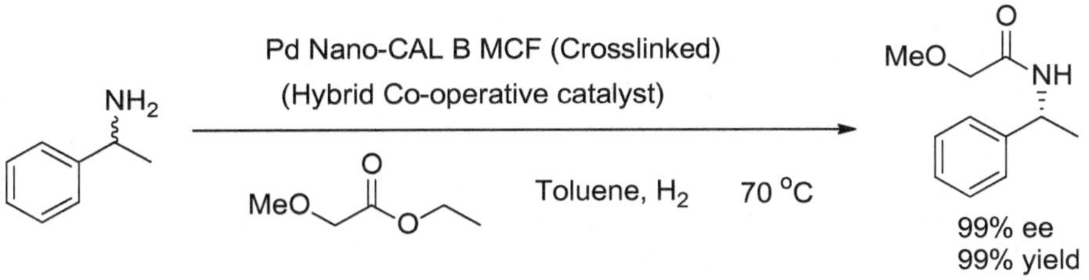

Scheme 4.14 Use of Pd nano-CALB co-cross-linked hybrid catalyst

The combination of organometallic catalyst and enzymes for DKR is specific and limited in number and scope. Finding a room temperature or below room temperature racemization catalyst compatible to the enzyme activity, with a high reracemization rate, requires a lot of efforts. Amines, mostly proved successful with benzylic systems, require high temperature. With the advent of nanotechnology hopefully, these difficulties can be overcome, and for almost all the enzymatic kinetic resolution, a suitable reracemization process would be available to the scientific community.

References

1. Patel RN (2000) Stereoselective biocatalysis. Mercel–Dekker, New York
2. Hoyos P, Pace V, Alcntara AR (2012) Dynamic kinetic resolution via hydrolase-metal combo catalysis in stereoselective synthesis of bioactive compounds. Adv Synth Catal 354:2585–2611
3. Huerta FF, Minidis ABE, Backvall JE (2001) Racemisation in asymmetric synthesis. Dynamic kinetic resolution and related processes in enzyme and metal catalysis. Chem Soc Rev 30:321–331
4. Martin Matute B, Bogger EK, Backval JE (2004) Highly compatible metal and enzyme catalysts for efficient dynamic kinetic resolution of alcohols at ambient temperature. Angew Chem Int Ed 43:6535–6539
5. Bogar K, Vidal PH, Alcntara Len AR, Backvall JE (2007) Chemoenzymatic dynamic kinetic resolution of allylic alcohols: a highly enantioselective route to acyloin acetates. Org Lett 9:3401–3434
6. Kazlauskas RJ, Weissfloch ANE (1997) A structure-based rationalization of the enantiopreference of subtilisin toward secondary alcohols and isosteric primary amines. J Mol Catal B: Enz 3:65–72
7. Allen JV, Williams JMJ (1996) Dynamic kinetic resolution with enzyme and palladium combinations. Tetrahedron Lett 37:1859–1862
8. Huerta FF, Laxmi YRS, Backvall JE (2000) Dynamic kinetic resolution of alphahydroxy acid esters. Org Lett 2:1037–1040
9. Kim MJ, Kim HM, Kim D, Ahn Y, Park J (2004) Dynamic kinetic resolution of secondary alcohols by enzyme–metal combinations in ionic liquids. Green Chem 6:471–474
10. (A) Parvulescu A, Vos DD, Jacobs P (2005) Efficient dynamic kinetic resolution of secondary amines with Pd on alkaline earth salts and a lipase. Chem Commun 42:5307–5309. (B) Reetz MT, Schimossek K (1996) Lipase-catalyzed dynamic kinetic resolution of chiral amines: use of palladium as the racemization catalyst. Chimia 50:668–669. (C) Choi YK, Kim MJ, Ahn Y, Kim MJ (2001) Lipase/palladium-catalyzed asymmetric transformations of ketoximes to optically active amines. Org Lett 3:4099–4101

11. Nielsen RJ, Goddard WJ (2006) Mechanism of the aerobic oxidation of alcohols by palladium complexes of N-heterocyclic carbenes. J Am Chem Soc 128(30):9651–9660
12. Choi YK, Suh JH, Lee D, Lim IT, Jung JY, Kim MJ (1999) Dynamic kinetic resolution of acyclic allylic acetates using lipase and palladium. J Org Chem 64:8423–8424
13. Kim MJ, Kim WH, Han K, Choi YK, Park J (2007) Dynamic kinetic resolution of primary amines with a recyclable Pd nanocatalyst for racemization. Org Lett 9(6):1157–1159
14. Engström K, Shakeri M, Bäckvall JE (2011) Dynamic kinetic resolution of β-Amino esters by a heterogeneous system of a palladium nanocatalyst and *Candida antarctica* lipase A. Eur J Org Chem 2011(10):1827–1830
15. Engström K, Johnston EV, Verho O, Gustafson KPJ, Shakeri M, Tai CW, Bckvall JE (2013) Co-immobilization of an enzyme and a metal into the compartments of mesoporous silica for cooperative tandem catalysis: an artificial metalloenzyme. Angew Chem Int Ed 52:14006–14010

Index

© The Author(s), under exclusive license to Springer Nature Switzerland AG 2023   45
A. B. Majumder and K. V. S. Ranganath, *Understanding Kinetic Resolution*
by Hydrolases, SpringerBriefs in Molecular Science,
https://doi.org/10.1007/978-3-031-46353-2